Foundations and Excavations in Decomposed Rock of the Piedmont Province

Edited by Ronald E. Smith

Geotechnical Special Publication No. 9

Foundations and Excavations in Decomposed Rock of the Piedmont Province

Proceedings of a session sponsored by the
Geotechnical Engineering Division of the
American Society of Civil Engineers
in conjunction with the ASCE Convention
in Atlantic City, New Jersey

April 28, 1987

Edited by Ronald E. Smith

Geotechnical Special Publication No. 9

Published by the
American Society of Civil Engineers
345 East 47th Street
New York, New York 10017-2398

ABSTRACT

The five papers in this volume deal with foundations and excavations in the Piedmont Province of the USA. This is a region underlain by very old metamorphic rocks that are overlain by residual soils that have been derived, in place, as a result of long term weathering processes. Between the surface soils and unweathered bedrock are varying stages of decomposed rock. The nature of this subsurface profile is highly variable, both horizontally and vertically. Each of the papers addresses one or more aspects of the design and/or construction of foundations and structures in this environment. The papers deal with some of the primary foundation types used in the Piedmont, namely shallow foundations, caissons and pressure injected footings. The cost of excavation and contractual aspects of caisson construction are also discussed.

Library of Congress Cataloging-in-Publication Data

Foundations and excavations in decomposed rock of the Piedmont Province.

(Geotechnical special publications; no. 9)
Includes indexes.
1. Foundations—Piedmont (U.S.: Region)—Design and construction—Congresses. 2. Excavations—Piedmont (U.S.: Region)—Congresses. 3. Engineering geology—Piedmont (U.S.: Region)—Congresses. I. Smith, Ronald E., 1938- II. American Society of Civil Engineers. Geotechnical Engineering Division. III. ASCE National Convention (1987: Atlantic City, N.J.) IV. Series.
TA775.F692   1987   624.1'52   87-1239
ISBN 0-87262-594-X

The Society is not responsible for any statements made or opinions expressed in its publications.

No part of this publication may be reproduced, stored in a retrieval system, or transmitted, in any form or by any means, electronic, mechanical, photocopying, recording, or otherwise, without prior written permission of the publisher.

TA
775
.F692
1987

Copyright © 1987 by the American Society of Civil Engineers,
All Rights Reserved.
Library of Congress Catalog Card No.: 87-1239
ISBN 0-87262-594-X
Manufactured in the United States of America.

FOREWORD

The Piedmont Province begins south of Atlanta and extends northeast beyond Philadelphia. Included in its boundaries are such cities as Wilmington, Baltimore, Washington, Richmond, Raleigh and Columbia. This geologic region is underlain by very old metamorphic rocks, predominantly gneisses and schists. These rocks have been subjected to millions of years of physical and chemical forces that have transformed the original rocks into various states of decomposition from relatively loose residual soils near the ground surface to hard, massive rocks at depth. The most consistent characteristic of the Piedmont subsurface profile is its variability both horizontally and vertically. At any specific location, variations result from such factors as: differences in original, premetamorphic parent materials; varying exposure to chemical and physical forces by way of complex rock discontinuity patterns; and intrusions and secondary mineralization within the rock mass.

As the geotechnical engineer embarks on a program of subsurface characterization for foundation purposes, he/she typically is faced with a significant degree of difficulty in obtaining high quality soil samples for laboratory testing. The difficulty of sampling is due to the presence of resistant core stones within the soil matrix. Even when the materials can reasonably be sampled, the performance of laboratory size samples often is not indicative of much larger scale ground mass behavior under actual building and excavation loadings. One of the principal reasons for this scaling problem is the presence of relict joints remaining within the residual soil/decomposed rock mass. As a result of these factors, the subsurface materials of the Piedmont typically have the worst behavior characteristics of both soil and rock.

The fact that the state of the practice of geotechnical engineering in the Piedmont has advanced in the face of formidable obstacles is largely a tribute to the pioneering efforts of Professor George F. Sowers of the Georgia Institute of Technology. Beginning in the mid 1950s, Professor Sowers started to publish the results of his findings relative to the engineering behavior of Piedmont materials. His continued efforts encouraged others to come forth and add other pieces to this complex geotechnical puzzle. As a result of past efforts, we can today deal with the variability and scale problems in reasonable ways.

It is the current practice of the Geotechnical Engineering Division that each paper published in a special publication be reviewed for its content and quality. These special publications are intended to reinforce the programs presented at convention sessions or specialty conferences and to contain papers that are timely or controversial to some extent. Ordinarily the reviews are carried out within a three-month period. The standards of review are essentially those for the *ASCE Journal of Geotechnical Engineering,* but the exigencies of timeliness and the need to have the publication available at the convention preclude more than one cycle of editing and revision. Therefore, it should be recognized that there are some differences in purpose between contributions to the special publications and those in the Journal. In accordance with ASCE policy, all papers published in this volume are open for discussion in the *ASCE Journal of Geotechnical Engineering,* and they are eligible for ASCE awards. Reviews of papers published in this volume were conducted by the Geotechnical Engineering Division's Committees on Rock

Mechanics and Engineering Geology in coordination with the Division Committee on Publications. The papers were reviewed by Paul W. Mayne, Carlos A. Mendez, Ronald E. Smith and Robert Waitkus.

We are indebted to the authors of the papers presented herein for adding to the body of knowledge. Each is a practitioner who lives with the reality of the Piedmont on a daily basis. What is presented can be applied to address foundation and excavation problems today and hopefully will help in advancing the state of the practice for tomorrow. Like all good engineering, these papers are aimed at providing cost effective solutions to real problems. It is the hope of the Rock Mechanics and Engineering Geology Committees that this volume will serve the current practice, and will encourage others, practitioners and researchers, to bring forward their findings to enhance the growth of the geotechnical profession.

> Ronald E. Smith
> Woodward-Clyde Consultants
> Rockville, Maryland
> February, 1987

CONTENTS

Settlement of Residual Soils
   R. E. Martin 1
Predicting the Difficulty and Cost of Excavation in the Piedmont
   R. M. White and T. L. Richardson 15
Pressure Injected Footings in Piedmont Profiles
   W. J. Neely, R. A. Waitkus and J. J. Schnabel 37
Design of Drilled Piers in the Atlantic Piedmont
   W. S. Gardner 62
Drilled Piers in the Piedmont—Minimizing Contractor-Engineer-Owner Conflicts
   S. A. Schwartz 87

Subject Index 103

Author Index 104

SETTLEMENT OF RESIDUAL SOILS

By Ray E. Martin[1], M. ASCE

ABSTRACT: The method previously suggested by the author in 1977 for calculating settlement of non-plastic to slightly plastic Piedmont Physiographic Region residual soils is reviewed. Additional pressuremeter data is included from throughout the Piedmont for development of a revised correlation between pressuremeter modulus and Standard Penetration Test N value. The rheological factor α relating the soil modulus of deformation to the pressuremeter modulus is re-evaluated based on new case histories and a value of 1 is again recommended. Thus, settlements may be calculated using the pressuremeter modulus as an approximation of the soil modulus of deformation when good quality pressuremeter modulus profiles are available for a specific site. A correction factor to be applied to the general settlement equation when rock occurs within the depth of influence of the foundation is presented. Settlements may also be calculated by estimation of pressuremeter moduli from N values. When this procedure is used, calculated settlements have been found to overestimate measured settlements and a correction factor of 0.6 is recommended to be applied to the general settlement equation.

INTRODUCTION

The writer (7) previously investigated the use of the pressuremeter modulus, E_{pm}, in conjunction with the Schmertmann (9) strain influence factor distribution for estimating settlement of non-plastic to slightly plastic residual soils of the Piedmont Physiographic Region. The method was developed by back calculation of the settlement of two foundation elements monitored during construction. Pressuremeter profiles were obtained below each foundation element prior to construction. The soil moduli of deformation, E_s, required to produce calculated settlements essentially equal to measured settlements, were determined and related to the pressuremeter moduli by a rheological factor α as recommended by Menard (8). A relationship was also developed for estimating the pressuremeter modulus, E_{pm}, from Standard Penetration Test (SPT) N values. The method has been found to overestimate measured settlement when N values are used to estimate the pressuremeter modulus, E_{pm}.

The purposes of this paper are fourfold. First, a wider geographic data base is desirable for development of the log E_{pm} versus log N

---

[1] Principal, Schnabel Engineering Assocs., P.C., One West Cary Street, Richmond, Virginia 23220

value relationship. Second, it is necessary to further evaluate the appropriateness of the assumption that $\alpha = 1$ for the type of residual soils under consideration. Third, a settlement correction for rock within the depth of the strain influence factor distribution is desirable. Fourth, a correction factor to the general settlement equation is required such that calculated settlements developed with E_{pm} values obtained from N value data more accurately estimate measured settlements. A total of nine case histories from the literature (3,4,6,7,13) have been evaluated in this analysis. The locations of the various case histories are indicated in Figure 1.

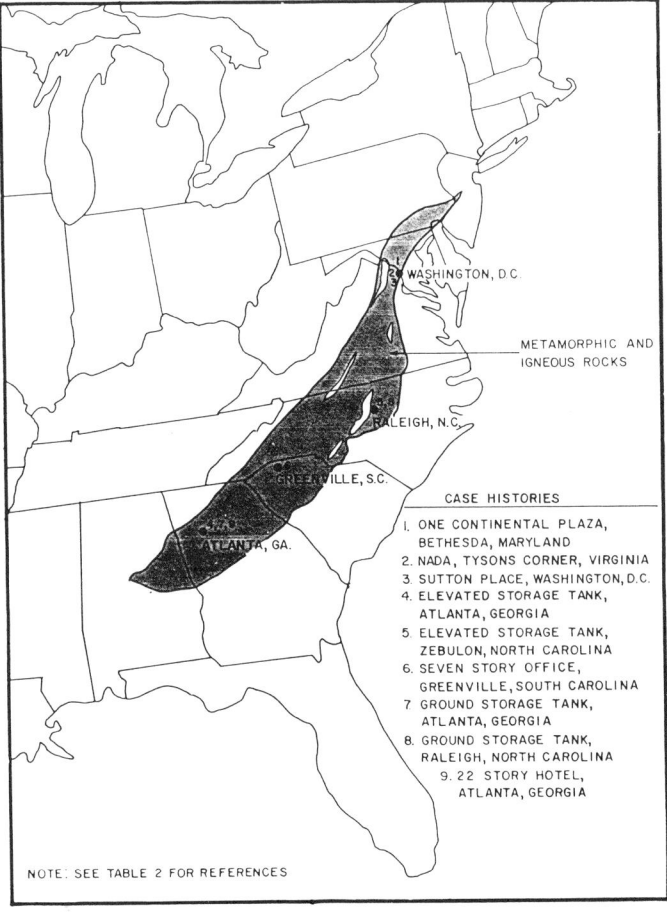

Figure 1 — Location of Case Histories

GEOLOGY

The original data used to develop the method included pressuremeter test results from the Washington, D.C. metropolitan area. The rocks from which the residual soil profiles were derived are typically gneisses and schists and granite intrusives. These soils classified SM and ML in accordance with the Unified Soil Classification System, ASTM D-2487 and are non-plastic to slightly plastic. The soils are also typically slightly to moderately preconsolidated based on consolidation test data. The preconsolidation effect occurs due to desiccation, removal of overburden, and relic mineral bonds.

In an effort to confirm the usefulness of the method to other geographic areas of similar Piedmont residual soils, additional pressuremeter data (3,4,6,11,14) has been added to the data base. The parent rock types remain the same. The general geographic region from New Jersey to Georgia where soils of this type exist in the Piedmont is indicated in Figure 1. The reader is referred to references 7 and 12 and to the introduction to this volume for a more detailed discussion of residual soil profiles.

CORRELATION OF E_{pm} AND N VALUES

The pressuremeter provides a practical solution to the problem of obtaining undisturbed samples for conventional laboratory testing in soils which are difficult to sample. Because this test is performed in situ, problems associated with sample disturbance which occur during pushing and extending thin walled tube samples are reduced. Tube samples of Piedmont residual soils are particularly vulnerable to this type of disturbance since they often contain rock fragments. This is true for soils where the N value exceeds about 10 to 15.

The pressuremeter was developed by Menard (8) and is widely used today in many types of soils. The test is performed by: (1) Preparing a cylindrical hole using either a 3 inch (76 mm) thin wall tube or a 3 inch (76mm) auger; (2) inserting the pressuremeter probe; and (3) measuring the volume displacement of the probe under increasing pressure steps. The data are plotted in the form of a volume versus pressure curve and the pressuremeter modulus, E_{pm}, is determined from the psuedo elastic phase. A typical pressuremeter test curve is included as Figure 2 illustrating the calculation of the pressuremeter modulus. The reader is referred to Baguelin et. al. (2) for detailed discussion of the pressuremeter test.

The original log E_{pm} versus log N value least squares regression analysis was performed using 120 data points for pressuremeter tests performed in the Washington D.C. metropolitan area (7). The least squares trend line for these data is designated line 1 in Figure 3. The correlation coefficient, r, was erroneously indicated to be 0.970 in reference 7. The actual correlation coefficient is 0.788.

During the past ten years, this data base has been expanded and includes data from the Virginia - Maryland area (11) within the

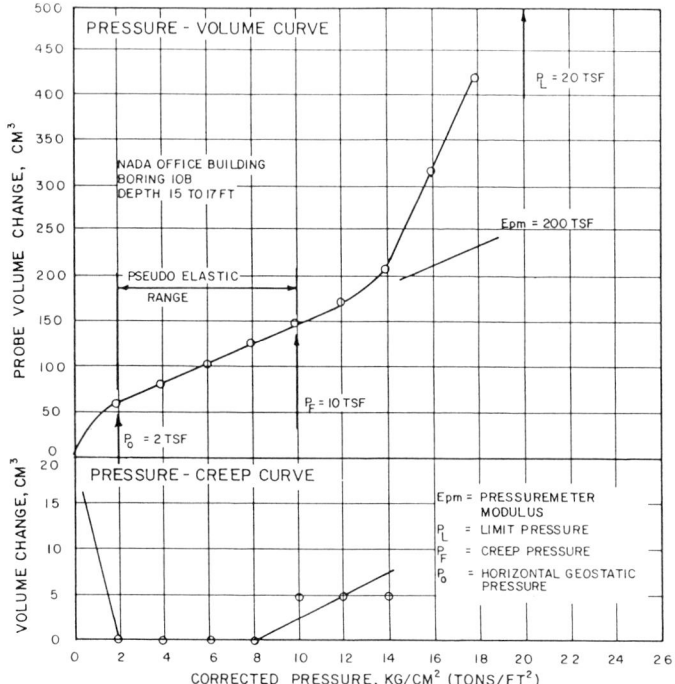

Figure 2 – Typical Pressuremeter Test Curves
(1 ft = 0.305 m, 1 tsf = 100 kPa)

Piedmont. The least squares trend line for the expanded data is designated line 2. The new correlation coefficient is 0.795 based on a total of 357 data points. An additional 102 data points were obtained from the literature (3,4,6,14). These data represent projects from Georgia to Maryland. When all 459 data points are considered, the new trend line designated line 3 falls slightly below line 2. The correlation coefficient is 0.790.

The validity of the linear relationship between log E_{pm} and log N value as defined by least squares trend line 3 was determined by using a t test of the correlation coefficient. The statistical relationship was found to be valid. The additional data included in the study have the effect of producing a more statistically correct relationship since the sample is larger and represents soils of a wider geographic region.

The scatter of the data in Figure 3 has caused some concern in the past for some users. Some of this scatter occurs due to variations in the soils tested. Due to the difficulty in defining soil profiles in Piedmont residual soils, no effort has been made to consider this type of variation. Another reason for the scatter is the quality of the actual pressuremeter tests. Considering this source of scatter, trend line 3 represents a conservative estimate of the actual value of E_{pm} which is desirable. Assuming the pressuremeter test is performed according to the standard procedures described in reference 2, the factors which effect the results most are an oversized borehole and disturbed borehole sidewalls. These factors will cause a reduction in the pressuremeter modulus, E_{pm}. Poor results from the SPT usually occur due to improper test procedures. Most of these variations from standard procedures, as described in ASTM D-1586, will produce increased N values. The individual or combined effect of these factors is to cause a data point to be plotted lower or farther to the right on Figure 3. Only in the less likely case of reduced N values due to disturbance will the data point move to the left and produce a less conservative result. Since the predicted E_{pm} values from Figure

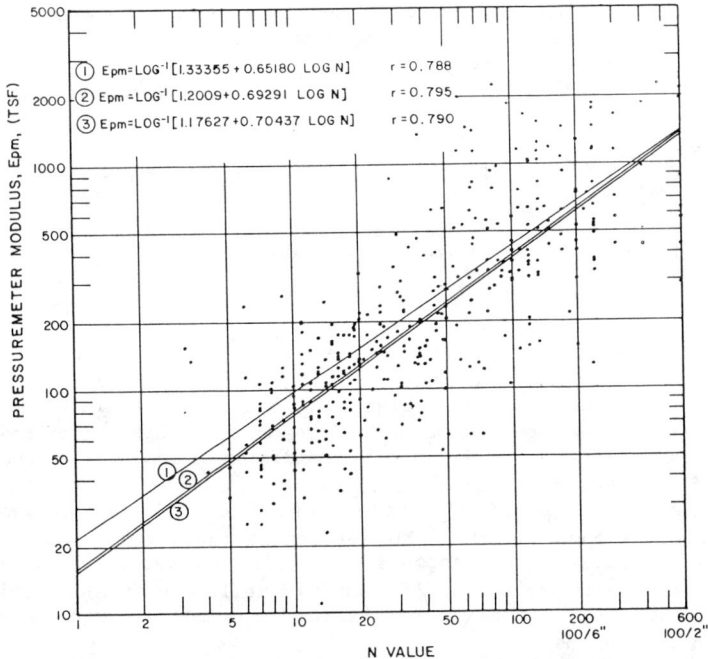

Figure 3 - Pressuremeter Modulus Versus N Value Data
(1 tsf = 100 kPa)

VERTICAL STRAIN DISTRIBUTION AND SETTLEMENT

The original strain influence factor distribution developed by Schmertmann (9) was modified by Schmertmann, et.al., (10) to include rectangular and square footings and a revised method for determining the peak strain influence factor as indicated in Figure 4. The general settlement equation is defined as

$$S = C_1 C_2 \Delta p \sum_0^{2B,4B} \frac{I_z}{E_s} \Delta z \quad \ldots \ldots (1)$$

in which C_1 is a correction factor related to foundation embedment; C_2 is a correction factor related to creep; Δp = net foundation pressure; I_z = vertical strain influence factor; and Δz = depth increment. The foundation embedment correction factor is defined as

$$C_1 = 1 - 0.5 \frac{p_o}{\Delta p} \quad \ldots \ldots (2)$$

in which p_o = the effective overburden pressure. The creep correction factor is defined as

$$C_2 = 1 + 0.2 \log \frac{t}{0.1} \quad \ldots \ldots (3)$$

in which t = the time in years since the load was applied.

The Schmertmann settlement analysis was developed considering elastic theory for a uniformly loaded area in conjunction with model tests of rigid footings, finite element studies and case histories. The method was developed for sands. Schmertmann indicated that the method could be used in evaluating the condition of rock within the depth of influence of idealized strain penetration. It is suggested that by assuming a sufficiently high modulus, such that the rock was essentially incompressible, the settlement predicted would be acceptable. This case corresponds to the elastic theory case of a rigid base occurring within the depth of influence of idealized strain penetration. Schmertmann indicated the method was applicable to rigid foundations.

Using the tables of influence factors developed from elastic theory provided by Harr (5) and Ahlvin and Ulery (1), it is possible to evaluate the effect a rigid base has on settlement below a uniformly loaded area. As the rigid base approaches a uniformly loaded area,

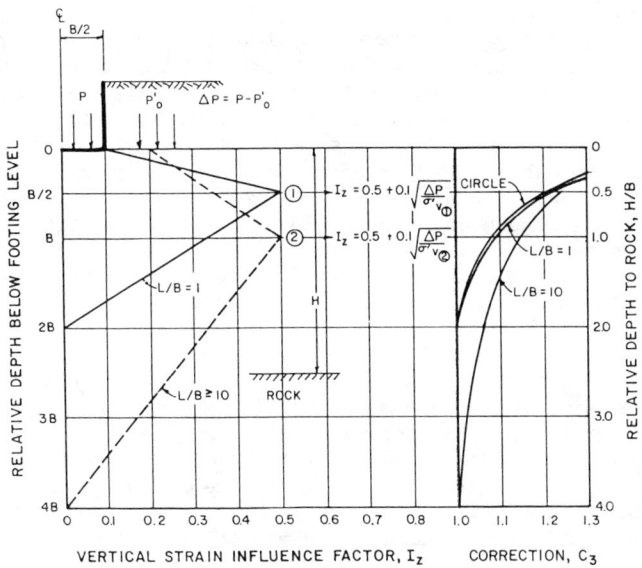

Figure 4 - Vertical Strain Influence Factor Distributions (10) and Correction C_3

for any point under the center line of the loaded area, the settlement is greater compared to the condition of no rigid base. It is possible to evaluate this increase in settlement by simply comparing the influence value for the rigid base at a particular depth below a loaded area with the influence value for the homogeneous isotropic elastic half space at the same depth.

The influence values for the homogeneous isotropic elastic half space were developed by subtracting the influence value at a particular depth, H, below the loaded area from the influence value for the surface. By dividing the influence value for the rigid base by the corresponding modified influence value for the homogeneous isotropic elastic half space, a correction factor is obtained. Since the recommended idealized strain influence factor distribution has been reduced to a finite depth by Schmertmann, it is logical to normalize the correction factor over depths of 2B for square footings (L/B=1) and uniform circular loadings and 4B for strip footings (L/B>10) where B and L are the foundation width and length, respectively. Figure 4 contains curves of a correction factor designated C_3 for the square, circular and rectangular footings assuming a Poisson's ratio of 0.3. This correction factor may be used to correct the general settlement equation.

The Schmertmann method is indicated to be applicable to rigid footings. By comparing influence values it can be determined that the difference in settlement associated with applying the same total load to a rigid footing or a flexible footing is usually less than 10 percent. The one case of practical interest is the circular uniform load such as for a ground storage tank. Settlements of several inches are typical and in this case the correction may have some significance. The correction can be made by comparing the influence values for a homogeneous isotropic elastic half space for both a circular uniformly loaded area and a square rigid footing. The correction to be applied to the general settlement equation is 1.06 for a circular uniformly loaded area.

CASE HISTORIES

A total of nine case histories included in the literature (3,4,6,13) were evaluated in the study including the two original case histories (7). The modified Schmertmann (10) strain influence factor distribution was used in these analyses. The case histories were selected based on adequacy of data related to loading, definition of subsurface conditions and settlement monitoring. Pertinent data concerning each case history are included in Table 1. The case histories included uniformly loaded tanks and square rigid footings, rock within the depth of influence of loading, and are widely distributed geographically as indicated in Figure 1. Settlements have been calculated for each case history and are compared to actual measured settlements. The corrections for rigid base and a circular uniform loading described above have been made in the computations as required by geometry.

A value $\alpha = 1$ was established in reference 7 for Piedmont type non-plastic to slightly plastic residual soils based on two foundation elements monitored. The case histories for which good quality pressuremeter modulus profiles are available were considered first to determine the appropriateness of the assumption $\alpha = 1$. For the original case histories (7), the NADA and One Continental Plaza office buildings, less than 0.1 inch change is indicated in the calculated settlement using the original and modified Schmertmann strain influence factor distribution. The measured settlements are about 90 percent of the new calculated settlements. The data are plotted in Figure 5.

Two additional case histories (4) where pressuremeter modulus profiles were available were considered. The E_s values for calculation of the settlement of a Ground Storage Tank and 22 Story Hotel in Atlanta, Ga. were estimated by assuming $\alpha = 1$. The measured settlements are about 70 and 85 percent of the calculated settlements for the tank and about 90 percent for the hotel as indicated in Figure 5. In the case of the tank, centerline settlement was calculated and corrected using the ratio of influence values for the centerline and edge condition for comparison to measured settlement at the tank

TABLE 1 - SUMMARY OF CASE HISTORY DATA

No.	Name	Location	Settlement Monitoring Location	Foundation Type	Footing Width (B) (ft)	Load (kips)	Average Bearing Pressure (psf)	Soil Type[3]	Relative Rock Depth H/B	Source of E_{pm}	Settlement (inches) Measured	Settlement (inches) Calculated	Settlement (inches) Corrected	Time t (yrs)	Reference
1	One Continental	Bethesda, MD	Footing 04.2	Rigid square	20	1292	3230	SM	1.15	PM profile Fig. 3	0.32 0.32	0.38 0.55	– 0.33	0.1 0.1	7
2	NVA Office	Tyson's Corner, VA	Mat #3	Rigid square	36	3071	2370	ML	>2	PM profile Fig. 3	1.1 1.3 1.1 1.3	1.22 1.47 2.28 2.74	– – 1.37 1.64	0.1 1.0 0.1 1.0	7
3	Sutton Place	Washington D.C.	Footing 15	Rigid square	10	468	4680	ML	>2	Fig. 3	0.42	1.09	0.65	0.1	11
4	Elevated Storage Tank	Atlanta, GA	Footing #6 Footing 110 Footing #2	Rigid[1] strip	11 11 11	37.5/LF 37.5/LF 37.5/LF	3410 3410 3410	ML,SM ML ML	1.27 1.40 0.91	Fig. 3 Fig. 3 Fig. 3	0.38 0.50 0.50 0.65 0.5 0.65	1.08 1.16 0.99 1.07 1.00 1.07	0.65 0.70 0.59 0.64 0.60 0.64	0.1 0.25 0.1 0.25 0.1 0.25	3
5	Elevated Storage Tank	Zebulon, NC	Footing 6/7 Riser Footing	Rigid square	13 16	633 1500	4000 6000	SM,ML ML	>2 >2	Fig. 3 Fig. 3	0.75 2.88	1.81 4.12	1.09 2.47	0.1 0.1	13
6	Seven Story Office	Greenville, SC	Footing G-14	Rigid square	19	1949	5400	ML	1.79	Fig. 3	2.25	4.08	2.45	0.1	6
7	Ground Storage Tank	Atlanta, GA	Bolts 2,3,4 Bolts 7,8,9	Uniform Circular	40 40	– –	4500 4500	SC ML	1.0 0.75	PM Profile[2] Fig 3 PM Profile[2] Fig 3	1.8 1.8 1.1 1.1	2.17 3.33 1.56 2.48	2.00 1.49	0.1 0.1 0.1 0.1	4
8	Ground Storage Tank	Raleigh, NC	Settlement Point P1 Settlement Point P3	Uniform Circular	120 120	– –	4065 4065	MH,SM ML	0.72	Fig 3 Fig 3	5.0 1.25	6.09 1.83	3.65 1.10	5.0 0.1	13
9	22 Story Hotel	Atlanta, GA	Boring B-5 Boring MB-1	Rigid strip	8.33 8.33	33.1/LF 33.1/LF	4000 4000	ML X	X	Fig 3 PM Profile Fig 3	1.1 0.8 0.8	1.48 0.88 1.21	0.89 0.73	0.1 0.1 0.1	4

1) Individual square footings (B=14.6 ft) at 18.25 ft on center assumed continuous based stress overlap analysis
2) Pressuremeter tests performed after tank load applied.
3) Unified Soil Classification System, ASTM D-2487.
4) 1 ft = 0.305m; 1kip = 4.45 kN; 1 psf = 992 kPa; 1 in. = 25.4 mm.

FOUNDATIONS AND EXCAVATIONS

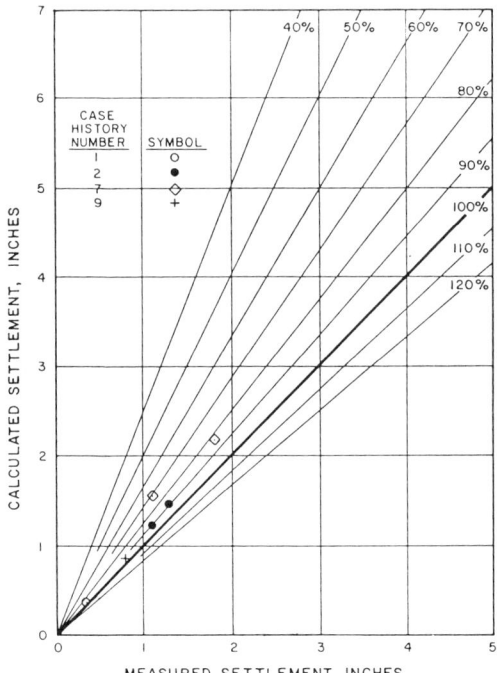

Figure 5 – Calculated Settlement Versus Measured Settlement From Pressuremeter Modulus Profiles (1 in. = 25.4 mm)

perimeter.

It should be noted, however, that the pressuremeter tests for the tank were performed adjacent to the tank after the tank had been loaded. The in situ stress conditions at the time the pressuremeter tests were performed were thus not the same as prior to loading. It would be anticipated that pressuremeter modulus values would be higher for these tests than if they had been performed prior to tank loading. Thus, the calculated settlements would be expected to be somewhat less than if pressuremeter moduli data had been developed prior to loading.

Based on the data presented in Figure 5 for the four case histories (4,7) in which pressuremeter modulus profiles were available at the location of the foundation element monitored, the original determination that $\alpha = 1$ appears reasonable.

Figure 3 was also used to estimate E_{pm} from N value data for all case histories (3,4,6,7,13). The factor $\alpha = 1$ was used and settlement computations were performed for each case history. Calculated

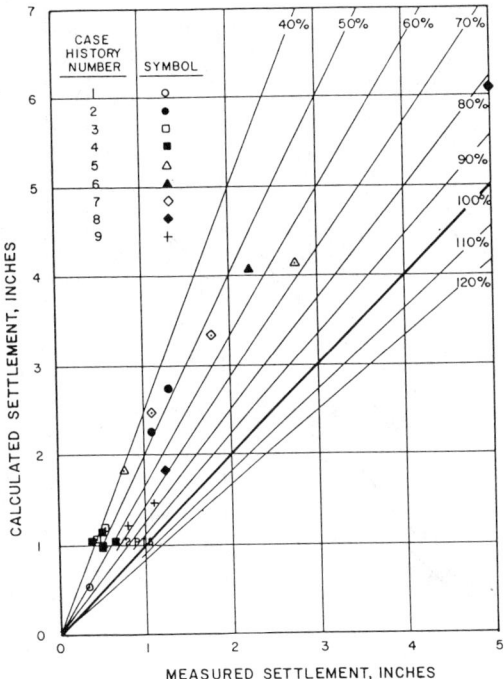

Figure 6 – Calculated Settlement Versus Measured Settlement For E_{pm} Estimated From Figure 3 (1 in. = 25.4 mm)

settlements are plotted versus measured settlements in Figure 6. A total of twenty points are indicated for fourteen different foundation elements. Multiple settlement readings were obtained on several of the foundations as indicated in Table 1. The measured settlements are indicated to be between about 40 and 80 percent of calculated settlements. Thus, a correction appears to be required to provide more reasonable settlement predictions when using the E_{pm} values obtained from Figure 3. A correction factor of 0.6 was applied to the previously calculated settlements and the data are replotted on Figure 7. The correlation between calculated and measured settlements is found to be more reasonable. For seventeen of the twenty points, measured settlements fall between 70 and 120 percent of calculated settlements. Thus, a factor 0.6 is recommended to be applied to the general settlement equation when the relationship in Figure 3 is used to estimate E_{pm}.

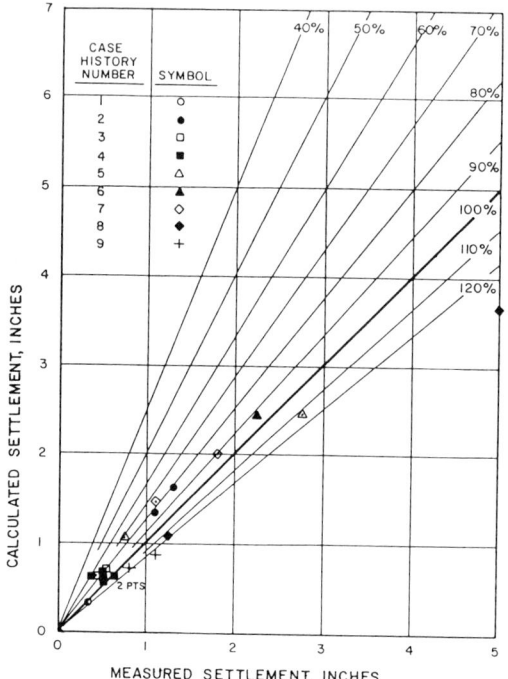

Figure 7 - Corrected Calculated Settlement Versus Measured Settlement For E_{pm} Estimated From Figure 3 (1 in. = 25.4 mm)

CONCLUSIONS

1. A revised correlation is presented in Figure 3 between log E_{pm} and log N for an expanded data base including 459 data points from the Piedmont Physiographic Region.

2. A correction factor, C_3, has been defined to account for the condition of rock within the depth of idealized strain influence. This factor is applied to settlements obtained from the general settlement equation. A correction factor of 1.06 should also be used for a circular uniform loading since the Schmertmann method was developed for rigid footings.

3. Based on the data available, the pressuremeter modulus, E_{pm}, may be used to approximate the soil modulus of deformation, E_s. The rheological factor α defined by Menard is thus 1. Soils for which this method applies include silty sands and sandy silts classified SM and ML which are non-plastic to slightly plastic and that are derived from weathering of metamorphic and igneous rocks.

4. Foundation settlements may be calculated by direct use of the

pressuremeter modulus in conjunction with the Schmertmann strain influence factor. The method presumes good quality pressuremeter tests are performed with minimal borehole disturbance and that a pressuremeter profile is developed.

5. Based on the case histories presented herein, the calculated settlement obtained by using Figure 3 to estimate E_{pm} and assuming $\alpha = 1$ will produce an overestimation of settlement. Thus, a correction of 0.6 to the general settlement equation is recommended.

ACKNOWLEDGEMENTS

The writer would like to thank Dr. Ron Smith, Woodward-Clyde Consultants for assistance in collecting data for the paper and for the overall coordination of the session volume. Thanks also go to Mr. Edward Drahos and Mr. Terry McGraw of Schnabel Engineering Associates, P.C., for assistance in reviewing the manuscript and performing the statistical analyses, respectively. I also gratefully acknowledge Schnabel Engineering Associates, P.C. for support of this effort through our in-house research and development program.

APPENDIX I – REFERENCES

1. Ahlvin, R. G., and Ulery, H. H., "Tabulated Values for Determining the Complete Pattern of Stresses, Strains, and Deflections Beneath a Uniform Circular Load on a Homogeneous Half Space", Highway Research Board Bulletin, No. 342, 1962, pp. 1-13.
2. Baguelin, F., Jezequel, J. F., and Shields, D. H., The Pressuremeter and Foundation Engineering, Transtech Publications, Rockport, Mass., 1978.
3. Barksdale, R. D., Bachus, R. C., and Calnan, M. B., "Settlements of a Tower on Residual Soil", Proceedings, Specialty Conference on Engineering and Construction in Tropical and Residual Soils, ASCE, Honolulu, Hawaii, 1982, pp. 647-664.
4. Barksdale, R. D., Ferry, C. T., and Lawrence, J. D., "Residual Soil Settlement from Pressuremeter Moduli", Proceedings, Uses of In Situ Tests in Geotechnical Engineering, ASCE, Geotechnical Special Publication No. 6, Blacksburg, Virginia, 1986, pp. 447-461.
5. Harr, M. E., Foundations of Theoretical Soil Mechanics, McGraw-Hill Book Co., Inc., New York, 1966, pp. 81-100.
6. Kahle, J. G., "Predicting Settlement in Piedmont Residual Soil with the Pressuremeter Test", Paper presented at the 1983 Transportation Research Board Meeting (unpublished), Washington, D.C.
7. Martin, R. E., "Estimating Foundation Settlements in Residual Soils", Journal of the Geotechnical Engineering Division, ASCE, Vol. 103, No. GT3, March 1977, pp. 197-212.
8. Menard, L., "Rules for the Calculation and Design of Foundations Elements on the Basis of Pressuremeter Investigations of the

Ground," translated by B. E. Hartman, Proceedings, 6th International Conference Soil Mechanics and Foundation Engineers, Montreal, Canada, Vol. II, 1965, pp. 295-299.
9. Schmertmann, J. H., "Static Cone to Compute Static Settlement over Sand", Journal of the Soil Mechanics and Foundations Division, ASCE, Vol. 96, No. SM3, May 1970, pp. 1011-1043.
10. Schmertmann, J. H., Hartman, J. P., and Brown, P. R., "Improved Strain Influence Factor Diagrams", Journal of Geotechnical Division, ASCE, Vol. 104, No. GT8, August 1978, pp. 1131-1135.
11. Schnabel Engineering Associates, P. C., "Menard Pressuremeter Results for Residual Soils in the Virginia-Maryland Area", (unpublished), 1986.
12. Sowers, G. F., "Residual Soils of the Piedmont and Blue Ridge", Transportation Research Board, Research Record, No. 919, 1983, pp. 10-16.
13. Willmer, J. L., Futrell, G. E., and Langfelder, J., "Settlement Predictions in Piedmont Residual Soil", Proceedings, Speciality Conference on Engineering and Construction in Tropical and Residual Soils, ASCE, Honolulu, Hawaii, 1982, pp. 629-646.
14. Woodward-Clyde Consultants, "Menard Pressuremeter Results for Residual Soils in Baltimore-Washington Area", (unpublished), 1986.

INVESTIGATION OF EXCAVATABILITY IN THE PIEDMONT

By Robert M. White[1] and Thomas L. Richardson[2], A.M. ASCE

Abstract: The lateral and vertical variability of rock weathering in the Piedmont presents significant problems in evaluating the probable cost of excavation for deep basements, pipelines, road cuts, subway excavations, etc. This paper presents information from a survey of geotechnical engineers and earthworks contractors who were contacted relative to methods of investigating excavatability of Piedmont materials. The paper also evaluates the type and amounts of information that would be cost effective in minimizing the cost uncertainties associated with the planning of excavations.

INTRODUCTION

The evaluation of the excavatability of weathered rock in the Piedmont and Blue Ridge geologic provinces of the eastern U.S. has proved to be a troublesome subject. Evaluation of excavatability is difficult primarily due to the residual profile in the Piedmont which yields a "transitional zone" between the overlying soil and underlying bedrock. This transitional zone has characteristics of both soil and rock, and can be erratic in consistency and extent. Attempts to describe its properties and to estimate excavation quantities have often led to claims and litigation. This is particularly true when persons unfamiliar with the nature of weathered rock in the Piedmont have attempted to describe it with more conventional engineering descriptions.

The Blue Ridge geologic province is adjacent to the Piedmont, and is similar geologically. We have included the two provinces in our discussions, and have collectively referred to them as the Piedmont.

The primary purpose of this paper is to evaluate the current practice with regard to the geotechnical investigation of excavatability in the Piedmont. This has been done by gathering information and opinions from owners, contractors and consultants knowledgable with excavation throughout the Piedmont and by characterizing their knowledge. This information has been combined with that of the authors to form recommendations to potentially improve evaluation of excavation difficulty and cost within the Piedmont.

The subject of this paper is the use of investigation methods to evaluate excavatability. This paper assumes the reader to have a basic knowledge of subsurface investigation methods. For that reason

1 Corporate Consultant, Law Engineering, Atlanta, GA 30328
2 Geotechnical Services Manager, Law/Geoconsult, Atlanta, GA 30328

we have not described these technologies in detail. A basic knowledge of excavation methods is also required to conduct a suitable excavatability investigation, and a more in-depth knowledge is required to estimate costs associated with excavation. However, these excavation methods are not the subject of the paper.

GEOLOGY OF THE PIEDMONT AND BLUE RIDGE

The Piedmont and Blue Ridge geologic provinces consist for the most part of metamorphic rock intruded by igneous rocks. Unmetamorphosed sedimentary rocks do occur within these provinces, however, they are not included as part of the subject of this paper. A figure in Appendix B gives the location of the Piedmont and Blue Ridge geologic provinces.

The Piedmont metamorphic rocks are predominantly metamorphosed sedimentary rocks typically of Precambrian and lower Paleozoic age and include gneisses, schists, amphibolites, phyllites, quartzite, slates and marble. Plutonic intrusions of Precambrian through Silurian age include gabbroic and granitic rock. Earlier plutons were deformed and/or metamorphosed by tectonic and metamorphic events during the Paleozoic resulting in their texture and engineering characteristics being similar to the metamorphic rocks around them. Later intrusives, on the other hand, typically retain their igneous texture. Diabase dikes were intruded during the Triassic.

Basins filled with clastic sediments, primarily sandstones, siltstones, and shales were formed during the Triassic. These unmetamorphized sedimentary rocks are not considered in this study because they are not similar to the crystalline metamorphic and igneous rocks typical of the Piedmont.

The Blue Ridge geologic province parallels the Piedmont geologic province to the northwest. In Alabama, Georgia and the Carolinas the Brevard fault forms the boundary between the Blue Ridge and the Piedmont. Like the Piedmont it consists primarily of metamorphic rocks intruded by igneous plutons.

Widespread development of joints (naturally occurring planar fractures in the rock) occurred within the Piedmont and Blue Ridge after the last major deformation, as a result of volume changes and directed stresses in the rock mass. Hydrothermal fluids moving in these joints often deposited mineral fillings. Common joint fillings include zeolite, calcite, chlorite, and quartz. The existence of jointing in the crystalline rock mass which tends to be open in the near surface allows the introduction of water and facilitates the weathering process. In many parts of the Piedmont, the distribution of vertical jointing controls the degree of weathering and therefore the development of topography. The degree to which a rock mass is jointed also strongly influences its excavatability.

Typical Piedmont and Blue Ridge rocks weather into a saprolite of variable thickness underlain by weathered and then fresh rock. Soil development tends to be deeper on ridge and hill tops and most shallow along streams and rivers yielding a rolling terrain. Drainage patterns tend to be rectilinear over foliated metamorphic rock and dendritic over igneous rock.

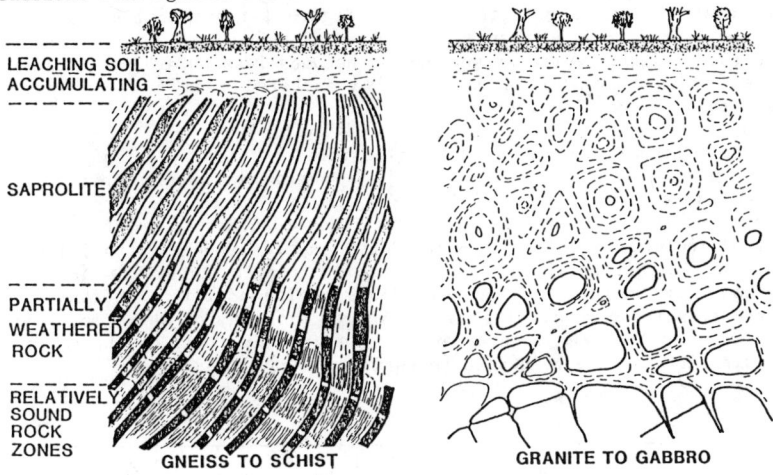

FIGURE 1 - Typical Piedmont Weathering Profiles
(after Sowers/Richardson, 1983)

Typical weathering profiles are shown in Figure 1. The weathering is most advanced at the ground surface and decreases with increasing depth. Four zones can be identified:

1. Upper zone - completely weathered with well-developed pedologic horizons,
2. Intermediate zone - saprolite with soil texture but retains relict structure of the original rock,
3. Partly weathered zone - alternate seams of saprolite and less-weathered rock, and
4. Unaltered or slightly weathered rock.

The boundaries of the zones are not well defined; the transition from one to another is usually gradual. The boundaries are usually not horizontal. The weathering is deeper and more advanced adjacent to fractures that transmit water and in mineral bands that are more susceptible to decomposition. Variations in thickness of residual soil of 20 m (75 ft) within a city block are not unusual. The variations are greatest in rocks that have steeply dipping foliation, or where adjacent parent rocks exhibit different resistances to weathering, or show pronounced variability in the extent of vertical jointing.

The partly weathered zone in strongly foliated rock (gneisses, schists, etc.) reflects the variability of weathering of the different rock bands. Weathering proceeds inward from joints through the less resistant rock producing deep soft seams alternating with seams of hard, relatively unweathered rock. In contrast weathering in granitic rock proceeds downward and inward from cracks more or less uniformly. The result is an XX pattern shown in Figure 1 that has boulder-like less weathered rock between the cracks. The boulders have an onion structure and the degree of weathering becomes less toward the boulder center. The thickness of the partly weathered zone can vary from negligible to over 100 feet.

The residual profile has been classified by some investigators by the degree of weathering [Figure 2]. Sowers has used a four-layer system since 1959 (Sowers, 1963). A similar system was proposed by Brecke for the Baltimore Region Rapid Transit Authority in about 1975 (Wirth/Ziegler, 1982). The Deere-Patton system (Deere/Patton, 1982) and the Law/Metropoliton Atlanta Rapid Transit Authority (MARTA) systems (Law Engineering report, 1980) provide additional definition of the transition zone. The major problem with any of these is their ability to define the boundaries in a useful way. All of these soil classifications have been used to estimate excavatability.

SOWERS 1963	DEERE & PATTON 1971		LAW/MARTA RICHARDSON/WHITE 1980	BRECKE 1975
SOIL $N = 5$ TO 50	I RESIDUAL SOIL	IA A-HORIZON	UPPER HORIZON - NO RESIDUAL STRUCTURE	RESIDUAL SOIL, RS
		IB B-HORIZON		
SAPROLITE $N = 5$ TO 50		IC C-HORIZON- SAPROLITE	SAPROLITE	RESIDUAL ZONE 1 $RZ-1$
PARTIALLY WEATHERED ROCK, ALTERNATE HARD & SOFT SEAMS $N > 50$	II WEATHERED ROCK	IIA TRANSITION SAPROLITE TO WEATHERED ROCK	PARTIALLY WEATHERED ROCK $N > 100$ CORE RECOVERY $< 50\%$	RESIDUAL ZONE 2 $RZ-2$
		IIB PARTLY WEATHERED ROCK	ROCK CORE RECOVERY $> 50\%$ RQD $< 50\%$	
ROCK RQD $> 75\%$	III UNWEATHERED ROCK RQD $> 75\%$		SOUND ROCK CORE RECOVERY $> 85\%$ RQD $> 50\%$	ROCK: Rx

FIGURE 2 - Classification of Piedmont Weathered Profile
(after Sowers/Richardson, 1983)

SURVEY PROCEDURES

A survey was conducted to assess industry practice in predicting the difficulty of excavations in the Piedmont. The survey was conducted by mailing a questionnaire to members of the engineering community involved in excavation. The questionnaire was developed by the authors and reviewed by Professor George F. Sowers. The questions are presented in the following section (along with a summary of the responses). The transmittal letter and explanatory material which accompanied the questionnaire are included as Appendix B.

A total of 65 questionnaires were sent to individuals representing owners, consultants, and contractors. The individuals were contacted by telephone prior to the mailing to ascertain their willingness to respond. Typically, they were also asked to supply the names of other individuals or organizations knowledgeable in the matter. In this manner, the base of contacts was expanded considerably. The potential respondants were contacted by telephone a second time several weeks after the questionnaire to encourage them to respond.

A total of 31 questionnaires were returned (48 percent of the total sent out). Table 1 indicates the responses divided among owners, consultants, and contractors. A list of the responding organizations is contained in Table 2 (Appendix A).

TABLE 1 – Summary of Responses

Category	Number of Questionnaires Sent Out	Number of Questionnaires Returned	Percentage
Consultants	29	16	55%
Owners	12	6	50%
Contractors	24	9	38%
Total	65	31	48%

SURVEY RESULTS

A detailed breakdown of the survey results is presented in Table 3 (Appendix A). A discussion of the results is presented herein. For clarity, we have presented the discussion on a question-by-question basis.

Question 1

What geotechnical investigation methods do you prefer in evaluating the excavatability of Piedmont weathered rock (borings with SPT, augering without SPT, coring, seismic refraction, percussion drilling)? If appropriate, indicate a combination of methods.

The most widely utilized investigative method was the standard soil test boring with SPT values. The recommended advantages of this method were that it provides standarized values recognized by most people involved in excavation and that it also provides additional information on soil consistency for use in other aspects of the project (such as fill placement). Also widely utilized was augering without SPT values. Economy was discussed as the major advantage of this method. A disadvantage is that it is dependent on the type of equipment and the experience of the driller, and that it yields qualitative, rather that quantiative, values. In addition, in Piedmont soil profiles the auger may refuse on small rock lenses as well as continuous rock and therefore provide misleading excavatability estimates.

Percussion (air track) drilling was noted as applicable in evaluating excavatability in areas of known high rock. It has the advantage of being able to describe rock below the refusal level of conventional drilling methods.

Seismic refraction was recommended by many respondents, particularly by consultants. It is used to provide information between borings and on linear excavations such as trenching. Most respondents described that where seismic refraction is used as the prime investigation method, it is usually correlated with borings or test pits. The boring or test pit information is often required in any event to provide information for building foundations or the suitability of the soil for fill. Several respondents responded that interpretation of seismic data requires an experienced and competent geophysicist. The limited applicability of seismic refraction in urban areas was noted.

Several respondents, particularly contractors but also consultants, recommended test pits utilizing a large backhoe, or trial excavations as the most direct means of assessing excavatability. This method is limited in depth and can be expensive.

No other geotechnical or geophysical methods, such as ground penetrating radar, cone penetration tests, seismic cross-hole, or resistivity, were considered appropriate by any of the respondents for excavatability investigations in the Piedmont.

Question 2

How well are you typically able to interpolate between borings (or seismic refraction lines)?

Opinions on interpolation between borings varied considerably. However, many stated that interpolation should be based on a general knowledge of the site, including geology and experience on nearby jobs. Several respondents reported successful interpolation required relatively closely spaced borings.

Question 3

What approximate spacing of borings (or seismic refraction lines) do you typically utilize? If you like, you may assume a typical site that you might be involved in. Would you recommend additional investigation effort to better define excavatability, taking into account the additional investigation cost?

Boring spacings of approximately 100 feet were widely recommended for approximately equi-dimensional excavation, such as for structures. Closer spacing, on the order of 50 feet, were strongly recommended in areas of anticipated difficult excavation. Seismic refraction between borings was recommended by some respondents in place of additional borings.

Consultants recommended that boring spacings of approximately 300 to 600 feet were generally appropriate for long excavations such as roads or utilities and trenches. Several consultants recommended primarily seismic refraction for this application. State DOT's recommended general boring spacing of approximately 200 feet for roads, with closer spacing in areas of anticipated difficult excavation.

Several contractors were strong in their recommendations that owners provide more subsurface information for excavation jobs. They commented that this would result in generally lower bids by the contractors.

Question 4

Do you find core recovery and RQD useful in evaluating excavatability in the Piedmont?

The question of the use of core recovery and RQD may have been poorly presented. Many respondents, particularly consultants, said they did not use this information. However, possibly implicit in this reply was that cored rock generally requires blasting and thus additional information was not required. Some respondents, particularly owners, said that the information was useful and some stated that a description of the rock core including degree of weathering, was as helpful or more helpful than recovery or RQD.

Question 5

Do you use the Caterpillar Tractor seismic velocity relationships in evaluating excavatability in the Piedmont?

Consultants for the most part used the Caterpillar tractor seismic velocity relationships while the owners and contractors tended not to. Consultants used lower velocity cutoffs for excavation estimates than does Caterpillar. These estimates may be influenced by the tendency of consultants to be conservative when estimating constructability.

Several consultants pointed out that the Caterpillar relationships were appropriate for open or flat excavation and that confined or sloping sites required a different relationship.

Owners in general did not make excavatability interpretations and therefore were unfamiliar with the Caterpillar relationships. One owner used the relationship but for in house slope design estimates.

Contractors were, for the most part, unfamiliar with the application of seismic refraction to excavatability studies.

Question 6

Based on investigation results, how do you subdivide the weathered rock material for predicting excavatability? What excavation methods will you envision utilizing in the subdivided zones?

Consultants used classification systems to interpret excavatability based on drilling and often seismic refraction data. Owners and contractors by in large did not. Standard penetration test N-values, seismic compression wave velocity and drilling core recovery were used by consultants but with wide variations in actual interpretations. For instance, one consultant felt that N=40 blows was the limit for conventional excavation while another preferred N=100/6". Most consultants viewed N=80 to 100 as the proper value. Similarly one consultant felt that a seismic compression wave velocity of 2000 feet per second (fps) was the limit for predicting conventional excavation while another chose 4000 fps. The average velocity chosen was 3000 fps. Similar differences occurred in using these parameters to predict the limit of rippability. Part of the variability may be related to the uneven incorporation of conservatism into the classification scheme by consultants rather than using the classification to predict excavatability (as did the Caterpillar Company) and then use conservatism. Another factor we suspect to be affecting our result is the relative expertise of the consultants. For both of these reasons numerical averaging of their results is inappropriate. Table 4 summarizes the classification approach taken by consultants.

TABLE 4 - Composite of Consultants Classification for Excavatability

Material	SPT "N"	Compression Wave Velocity "Vp" fps	Excavation Technique
Soil/ Weathered Rock Boundary	40 - 100/6" 80 - 100 typical	2500 - 4500 3500 typical	Boundary between conventional and ripping
Weathered Rock/ Rock Boundary	100 - refusal 100/4" typical	4000 - 8000 6000 typical	Boundary between ripping and shooting

EXCAVATABILITY INVESTIGATION

Contractors typically provided either general descriptions of materials (saprolite, partially weathered rock, etc.) with anticipated excavation methods, or simply described the materials by their excavatability (rip rock, blast rock, etc.). Limited references to excavation methods were included.

Owners either did not classify materials or provided responses similar to contractors.

Questions 7 and 8

7. Are there specific weathered rock conditions in the Piedmont for which excavatability is particularly difficult to evaluate? Please describe. Do these conditions often lead to claims? If appropriate, attach case history information (in any form) which involved problems in estimating excavatability.

8. Referring to the previous question does it appear feasible to improve the evaluation of these conditions with standard or special investigation techniques? Are the conditions specially related to various areas within the Piedmont?

All groups recognized boulders as a weathered rock condition which was difficult to evaluate. The consultants recommended seismic refraction combined with borings or test pits. The thinking expressed by one consultant was that if borings encountered a high refusal level and seismic refraction indicates a low velocity for underlying material, then bouldery conditions can be inferred. One owner recommended percussion drilling or coring to evaluate bouldery conditions. The contractors had no recommendation.

Extreme differential weathering was recognized by the consultants and contractors. In general, no special methods were suggested to improve investigation of this condition. One contractor suggested that exploration be managed by an experienced engineering geologist so that all available geologic factors could be used.

Question 9

Could you provide us with a range of unit costs for various excavation methods in Piedmont rock, with some comment as to how you evaluate these costs?

The responses to this question showed general agreement on excavation costs and respondents pointed out factors which influence excavation costs, namely open versus trench or confined excavation and large versus small jobs. Most respondents reported inclusive costs but some obviously did not, and as few respondents gave detailed examples of estimating, we could not convert the partial cost to complete costs. The only owners to respond were state highway departments and as they all used unclassified bidding they were unable to provide costs by material type. One contractor had very high costs and mentioned that when he used these costs to bid he never got the job! Consultant costs were 30-50% higher than contractor costs. A possible explanation is that most contractors specified large excavations (100,000 c.y.) and the consultants did not specify size.

Question 10

> Would you give us an example of your estimating on an excavation job using the results of the exploratory technique that you most commonly use?

There was not sufficient response to this question to evaluate the results.

Question 11

> How much do you typically spend on an excavatability investigation? If possible, express this as a percentage of construction (excavation) cost. How does this differ for open versus trench excavation?

Consultants and contractors generally responded to this question. Only one owner, a state highway department, responded.

Consultants and contractors both estimated actual exploration costs to be less than or equal to 1% of excavation costs for open excavations. Contractors and consultants felt exploration for trenches was more expensive, 2% or more. Consultants also expressed their costs as a function of area and length. Also some consultants gave flat costs without regard to any parameter. It is not understood how these costs are determined.

Contractors made the point that their exploration was prebid and more exploration may be undertaken by them after the job is secured to effectively plan the work. Contractors also stated that they often spent nothing on exploration if the job was of low interest to them.

The only owner to respond to this question, a state highway department engineer, estimated $6000 per highway mile.

Question 12

> How do groundwater considerations enter into your excavatability and cost considerations?

Two-thirds of the respondents felt that groundwater must be considered and at times could be significant for cost estimates. One third felt that groundwater was either not significant or not usually significant. Consultants were split on their responses, while contractors felt strongly that the case of the subgrade in weathered rock was particularly sensitive to groundwater, because a wet working area could weaken the subgrade and slow excavation.

DISCUSSION OF RESPONSES

Bouldery conditions and erratic depths of weathering were commonly identified as causing difficulties in predicting excavatability. Erratic depths of weathering in the Piedmont are typically related to the following geologic conditions:

1) Variable spacing of open high angle joints;
2) Variable bedrock lithology; and
3) Steeply dipping rock formations.

Bouldery conditions are typically related to weathering of unfoliated igneous rock.

The responses from the consultants, contractors and owners showed different points of view and knowledge that the groups have developed in response to their different roles in excavation projects.

Consultants appear more knowledgeable concerning exploration than the other groups. However, contractors know more about the technology of excavation than many of the consultants. Consultants base their exploration on material classification systems which allow them to characterize the subsurface and estimate excavatability from drilling, seismic and other data.

Contractors want information rather than data and when they explore they tend to use techniques directly related to excavatability such as test trenches and percussion drilling using their own equipment. Consultants are more confortable using conventional drilling and seismic refraction which they are often in the business of providing.

Large owners (utility and state highway departments) often have their own personnel and equipment for exploration while smaller owners often hire consultants. Contractors rarely hire consultants in their routine exploration and cost estimating to prepare bids.

Owner's excavation costs on a given job are based on what they pay for the work and not what it costs to do. By providing some information and allowing competitive bidding they pay the lowest of the cost estimates made by the competing contractors. Owners vary as to whether they use unclassified or classified bidding.

Contractors costs as given to us were values used in bids and reflect estimated costs, profit, uncertainty and how badly the contractor wants the job. Consultants are not directly impacted by actual excavation costs and their cost estimates tend to be conservative because they are more comfortable "safe than sorry", safe meaning over estimating excavation costs.

SUGGESTED APPROACHES TO EXPLORING PIEDMONT SITES FOR EXCAVABILITY

Several types of subsurface conditions have been identified which are difficult to evaluate and thus most likely to produce unpredictable excavatability. These include bouldery conditions and erratic depths of weathering. Yet, both of these conditions are related to the nature of the parent rock. Thus, increased knowledge of geologic conditions should yield a better grasp of the horizontal and vertical extent of the weathered profile. However, geologic interpretation is often difficult because of subtle distinctions between lithologies, unpredictable extent of the varied rock types, and because the weathering profile often leaves only an indirect record of the parent rock. However, a general understanding of the site geology should be the first step in assessing excavatability. The geologic assessment should be based on examination of outcrops, available geologic information, and rock core if available. Careful examination of residual soil samples can be indicative of the parent rock type. If outcrops are associated with past excavation, construction case histories should be secured if possible.

It is the authors' opinion that the methods recommended by the respondents are appropriate for use in evaluating Piedmont excavatability, and we do not find appropriate other potential methods such as ground penetrating radar or cone penetration tests. The investigation methods should be carefully planned by the owner and/or his consultant. The various investigative methods should be evaluated, with the advantages and disadvantages weighed. It appears that the conventional soil test boring with Standard Penetration Tests will remain the main investigative method, due primarily to its wide acceptance. However, the other methods can provide valuable information and should be considered for specific applications. Auger boring without standard penetration testing will probably continue in use due to its economy. However, its applicability is limited because the results are dependent on the characteristics of the equipment and operator. In addition, augers without sampling provided limited information for purposes other than excavation. Soil borings either with or without sampling can, in some cases, be misleading where refusal is encountered on a rock lens or boulder. Rock coring solves this problem but is often not used due to cost. Seismic refraction is valuable for several reasons. One is its ability to assess large areas at relatively low cost. This is particularly appropriate for preliminary investigations and for long narrow excavation, such as utility trenches. Seismic refraction is well suited to assess bouldery conditions which cannot be properly investigated with borings. Test trenches undoubtedly yield the best "hands-on" feel for excavatability and should certainly be used when feasible. Often, a combination of methods will provide the best results.

Additional investigation costs are worthwhile where excavation costs represent a significant portion of the project and where unit costs are part of the contract, or where delays due to additional excavation effort could impact the project. The additional effort should be targeted toward the variable conditions identified previously. Additional investigation effort may, in some cases, be best accomplished with extra borings, seismic refraction, or test trenches.

REFERENCES

1. Deere, D.U. and Patton, F.D., "Slope Stability in Residual Soils", Proc., 4th Pan American Conference on Soil Mechanics and Foundation Engineering, San Juan, 1971

2. Law Engineering Testing Company, Report of Subsurface Investigation, Final Design, DN-430, Metropolitan Rapid Transit System, Marietta, Georgia, May 1980

3. Sowers, G.F., "Engineering Properties of Residual Soils Derived from Igneous and Metamorphic Rocks", Proc., 2nd Pan American Conference on Soil Mechanics and Foundation Engineering, Brazil, 1963

4. Sowers, G.F. and Richardson, T.L., "Residual Soils of the Piedmont and Blue Ridge", Transportation Research Record 919, p.10-16, 1983

5. Wirth, J.L. and Zeigler, E.J., "Residual Soils Experience on the Baltimore Subway", ASCE Specialty Conference on Engineering and Construction in Tropical and Residual Soils, Honolulu, 1982

SUPPORTING DOCUMENTS SUBMITTED BY RESPONDENTS

1. Church, H.K., *Excavation Handbook*, McGraw Hill, 1975.

2. Obermeier, S.F., "Engineering Geology of Soils and Weathered Rocks of Fairfax County, Virginia" U.S. Geologic Survey Open-File Report 79-1221, 1979

3. Woodward Clyde Consultants, report for client (Montgomery County, Maryland), 1980

FOUNDATIONS AND EXCAVATIONS

APPENDIX A - TABLES 2 AND 3

TABLE 2 — List of Respondents

The following companies or agencies respondend to our questionnaire. Their efforts are greatly appreciated.

CONSULTANTS

ATEC Associates
° Columbia, Maryland
° Raleigh, N. Carolina

Professor Wayne Clough
Virginia Polytechnical Institute
Blacksburg, Virginia

Dames and Moore
Bethesda, Maryland

Fisher Associates
Durham, N. Carolina

Jenny Engineering Corp.
Springfield, N. Jersey

Law Engineering Testing Company
° Atlanta, Georgia
° Charlotte, N. Carolina
° Washington, D. C.

Parsons, Brinckerhoff Quade & Douglas
Atlanta, Georgia

J. N. Pease Associates
Charlotte, N. Carolina

Rummel, Klepper & Kahl
Baltimore, Maryland

Schnabel Engineering Associates
Bethesda, Maryland

Soils and Materials Engineers, Inc.
° Atlanta, Georgia
° Raleigh, N. Carolina

Woodward-Clyde Consultants
Rockville, Maryland

OWNERS

Georgia Dept. of Transportation

Maryland State Highway Administration

N. Carolina Dept. of Transportation

S. Carolina Dept. of Highways and Public Transportation

Southern Company
Birmingham, Alabama

Washington Metropolitan Area Transit Authority

CONTRACTORS

APAC Georgia, Inc.
Atlanta, Georgia

Barnhill Contracting Company
Atlanta, Georgia

Blount Construction Company
Atlanta, Georgia

Brakefield Construction Company, Inc.
Casey, S. Carolina

Chatham Brothers, Inc.
Atlanta, Georgia

D. W. Flowe & Son
Charlotte, N. Carolina

Harp Grading Company, Inc.
Atlanta, Georgia

Nello L. Teer Company
Durham, N. Carolina

U. S. Construction Company
Columbia, S. Carolina

TABLE 3

Detailed Summary of Responses to Questions

CONSULTANTS

1.

16 of 16 provided responses.
All respondents mentioned conventional soil borings with SPT values.
Several (4) mentioned that considerable information can be gained from borings without SPT's performed by an experienced driller.
Most (11) mentioned use of seismic refraction in combination with test borings, some noting that seismic refraction was particularly applicable in areas of difficult access.
Several (4) mentioned that the best method was a test pit excavated with a large backhoe or a test excavation for rippability (depending on the proposed depth of excavation).
Several (6) mentioned rock coring.
Many (8) mentioned that a combination of methods often yielded the most reliable results.

2.

16 of 16 provided responses.
Most (10) agreed that interpolation often yielded problems.
Several (5) noted that interpolation should be tied to knowledge of geology with conditions such as steeply dipping rock or dikes and sills making interpolation particularly difficult.
Several (3) noted that seismic aids in interpolation.

OWNERS

1.

6 of 6 provided responses.
All respondents preferred borings, some (2) preferring augering without SPT, some (4) preferring SPT's.
All mentioned coring.
Several (3) respondents mentioned seismic refraction, one noting its limitations in urban areas.
Two mentioned percussion drilling, one for application in areas of steep terrain or shallow rock, one for use with down-hole seismic.
One mentioned combination of methods should be used.

2.

6 of 6 provided responses.
Some cited good results, others cited poor results.
One noted that good results based on knowledge of geology.
One said good results based on performance of extra borings in questionable areas.
One said top of rock requiring blasting more predictable top of material requiring ripping.

CONTRACTORS

1.

9 of 9 provided responses.
Most (6) preferred borings with SPT.
Some (3) preferred borings without SPT.
Several (2) preferred backhoe excavated test pits.
One utilized seismic refraction in combination with borings and test pits.
One utilized percussion drilling where needed.

2.

9 of 9 provided responses.
Several (2) respondents tied their interpolation success to the closeness of the boring spacing.
One said interpolation depended on knowledge of rock types.
One cited interpolation problems in the "fringe" area of Piedmont.
One cited need to contour top of rock as aid to interpolation.

TABLE 3
Detailed Summary of Responses to Questions
(continued)

CONSULTANTS

3.

16 of 16 provided responses. Most cited boring spacing on the order of 100' for buildings or other approximately equi-dimensional excavation. Estimated ranged from 50 to 200 with some citing geology as determining actual spacing. Several recommended the closer spacing when difficult excavation was anticipated.
For long routes, such as roads or utilities, boring spacing of 300 to 600 feet were recommended. Several recommended primarily seismic refraction confirmed with borings or test pits.

4.

16 of 16 provided responses. Most (9) stated that core recovery and RQD were not helpful. However, probably implied in this response (and stated in some) was that all cored material requires blasting, so that this infomaton was primarily of interest to the blasting engineer.
Six said core recovery and RQD were useful.
One stated that coring was useful in the Brevard zone where cored material is often rippable.
Two said a description of rock weathering was as good or better than recovery or RQD.

OWNERS

3.

5 of 6 provided responses.
For roads, several recommended a boring spacing of 200 feet supplemented with additional borings or seismic refraction required.
Two mentioned closer spacing in areas of difficult geology.
One recommended auger borings (without SPT) or percussion holes at 100 foot spacing.

4.

5 of 6 responded.
All (5) said that core recovery and RQD were useful.
Two said that recovery depended on driller's skill.

CONTRACTORS

3.

9 of 9 provided responses.
Most recommended 50 to 100 foot spacing.
Several recommended wider (200 feet) spacing where rock is not anticipated and closer spacing (50 feet or less) in areas where rock is anticipated.
Several stated that boring information provided by owners was usually inadequate and that a through investigation more than pays for itself.

4.

9 of 9 provided responses.
Approximately half said this information was useful.

EXCAVATABILITY INVESTIGATION

TABLE 3

Detailed Summary of Responses to Questions
(continued)

CONSULTANTS	OWNERS	CONTRACTORS
5.	5.	5.
13 of 16 provided responses. Most (11) respondents used the Caterpillar Tractor relationships with the following reservations: • the values are optimistic or do not contain sufficient conservatism for a consultant (4 respondents). • they apply specifically to open or flat excavation (13 respondents). • use as a guide along with other data (3 respondents).	5 of 6 provided responses. Most respondents (4) did not use the relationships. • One respondent report velocity values to bidders but does not interpret excavatability. • One felt the relationships were not accurate enough, only for in-house slope design estimates.	7 of 9 provided responses. Most (5) did not use the relationships. Two used the relationships with other data, such as borings and a geologist to interpret.
6.	6.	6.
15 of 16 provided resposes. Most respondents (13) provided correlations between investigation results and excavatability. Many were comprehensive and detailed. See Table 5 for a summary.	5 of 6 provided responses. Two respondents did not subdivide the weathered rock material for predicting excavatability. The remaining three respondents provide only general and limited subdivisions based on excavation methods.	8 of 9 provided responses. All respondents provided at least limited subdivisions bt they were mostly described in ecxcavation terms rather than in terms of investigation results. Some respondents (6) included descriptions of materials in conventional terms (saprolite, partially weathered rock, etc.) but did not relate these to investigation results.
7.	7.	7.
9 of 16 provided responses. Most (5) thought bouldery conditions were difficult to evaluate. Most (5) also thought extreme variability in the weathering profile was also difficulr to evaluate.	2 of 6 provided responses. Both respondents thought bouldery conditions were difficult to evaluate. One respondent thought sloping rock formations were difficult to evaluate.	4 of 9 provided responses. Each of the respondnets had a different condition: • boulders • vertical rock formations • mud and rock • variable lithology (missed by boring pattern).

TABLE 3
Detailed Summary of Responses to Questions
(continued)

CONSULTANTS	OWNERS	CONTRACTORS
8.	8.	8.
6 of 16 provided responses. Five respondents thought that seismic refraction combined with borings or test pits was the best method to evaluate bouldery conditions. One respondent thought it was not possible to improve investigation of these conditions. No special methods were suggested to evaluate highly variable conditions.	1 of 6 provided a response. The respondent suggested percussion drilling supplemented by coring to evaluate bouldery conditions.	3 of 9 provided responses. Each of the respondents suggested a different improvement: ° test excavations ° experienced engineering geologist managing investigation (for variable lithology) ° more rock coring. Two respondents thought it was not possible to improve methods.
9.	9.	9.
6 of 16 provided responses. soil. . . . . $ 3 - 6 /c.y. ripping. . . . . . $ 5 - 15 /c.y. blasting . . . . . $10 - 40 /c.y. (open excavation) $30 - 60 /c.y. (trench) $60 - 500 /c.y. (footing)	2 of 6 provided responses. Unclassified excavation $1.40-$2.80 /c.y. (large volumes). We interpret this to be a partial cost.	7 of 9 provided responses. soil . . . $2-4/c.y. ($2/c.y. typical) (one at $10/c.y.) ripping . . $2.5-10/c.y. ($3.5/c.y. typical) (one at $75/c.y.) blasting . open $18-32/c.y. (one at $100/c.y.) trench $80-100/c.y. (above are large jobs > 1000,000 c.y.)
10.	10.	10.
Insufficient response.	Insufficient response.	Insufficient response.
11.	11.	11.
6 of 16 provided responses. Three types of responses: 1) By acre or foot Open Excavation $ 50-350 /ac. Trench Excavation $0.30-0.50/ft. 2) As percent of excavation: Open Excavation $\leq$ 1% 3) Flat costs: $500 - $10,000	1 of 6 provided a response. $6000/mile for highway construction.	6 of 9 provided responses. Open excavation - typically $<$ 1% large jobs 0.2% Trench excavation = 2%+ note: 1) pre-bid costs, more investigation later 2) some contractors investigate only selected jobs.

32 FOUNDATIONS AND EXCAVATIONS

TABLE 3

Detailed Summary of Responses to Questions
(continued)

CONSULTANTS	OWNERS	CONTRACTORS
12.	12.	12.
10 of 16 provided responses. The responses were approximately split: Six thought groundwater must be considered and can be significant. Four thought groundwater considerations were not very significant.	3 of 6 provided responses. Two thought groundwater considerations were not significant. One thought that it could weaken subgrade in weathered rock.	9 of 9 provided responses. Most (8) thought groundwater must be considered and can be significant, particularly in operation sequencing.

APPENDIX B- TRANSMITTAL LETTER AND EXPLANATORY LETTER WHICH ACCOMPANIED QUESTIONNAIRE

November 20, 1986

....
....
....

Attention: ....

Dear ....:

Thank you for agreeing to provide information for our paper on excavation in the Piedmont and Blue Ridge. As I explained on the telephone, we have been asked to prepare a paper on this subject for presentation at the spring ASCE convention in Atlantic City. The paper will be part of a session dealing with foundations and excavations in the Atlantic Piedmont. We have attached an abstract which more fully explains our proposed paper.

In the southeast, the metamorphic and igneous rock between the Coastal Plain sediments and the Valley and Ridge rocks are referred to as the Piedmont or Blue Ridge. North of approximately Maryland, this area of rock thins and the different rock types are locally referred to as the Wissahickon Schist, granite gneiss, or other local variations. We are interested in your experience in these metamorphic and igneous rocks.

We request that you fill out the attached questionaire. If you would like to provide additional information, please feel free to attach it in any convenient form. If you do not feel qualified to respond to any of the questions, leave them blank. As a convenience we have attached a plan showing the extent of the Piedmont and Blue Ridge, and a general description of Piedmont subsurface materials indicating our zone of interest.

The information we gather will be summarized and presented as part of the paper. We intend to list all respondents within our paper. In addition, we will be glad to send you a preprint of the paper when it becomes available.

Naturally, we need this information yesterday. It would be greatly appreciated if you could fill out and return the questionaire in the next week or two.

Very truly yours,

Thomas L. Richardson, P.E.   Robert M. White, P.E., P.G.
Law/Geoconsult International   Law Engineering Testing Company

TLR/Mcph.
attachments.

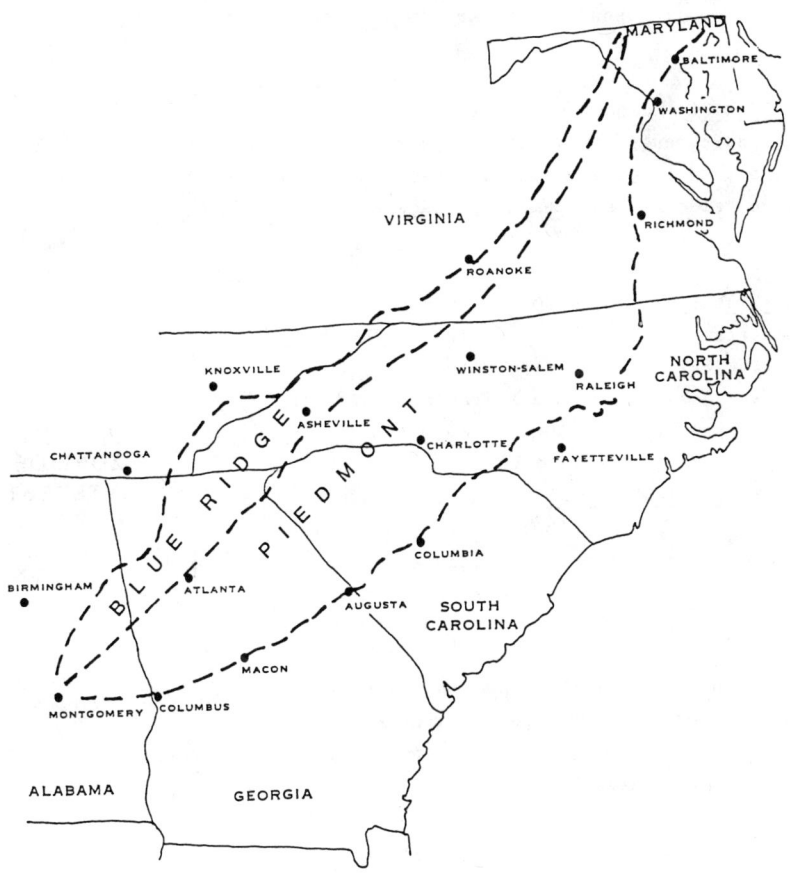

General Description of
Subsurface Materials and Zone of Interest

SAPROLITE – decomposed rock (soil) with standard penetration test (SPT) resistances less than 100 blows/foot and seismic refraction velocity less than approximately 3000 fps.

PARTIALLY WEATHERED ROCK – decomposed rock penetrated with soil investigation methods having SPT resistances greater than 100 blows/foot and seismic refraction velocities from approximately 3000 to 6000 fps.

ROCK (FRACTURED ROCK) – investigated with rock coring methods, exhibiting frequent breaks and weathered zones; core recoveries generally less than 75%, seismic refraction velocities from approximately 6000 to 10,000 fps.

} **PRIMARY ZONE OF INTEREST**

SOUND ROCK – generally unweathered rock, core recoveries generally greater than 85%, seismic refraction velocities greater than approximately 10,000 fps.

PRESSURE INJECTED FOOTINGS IN PIEDMONT PROFILES

William J. Neely,[1] M.ASCE, Robert A. Waitkus,[2] M.ASCE, and James J. Schnabel,[3] M.ASCE.

INTRODUCTION

Pressure injected footings (PIF), also referred to as Franki Piles or compacted concrete piles, are widely used for the foundations of structures throughout the Piedmont physiographic province. Since 1968 more than 100 projects have been completed with typical pile lengths ranging from about 15 to 30 ft. Design capacities for pressure injected footings have increased steadily from about 70 tons to as much as 200 tons on some recent projects.

An analysis of data from almost 100 projects completed in the Piedmont province since 1968 shows that the mean PIF capacity increased from about 100 tons in 1968 to almost 170 tons in 1986. On individual projects, maximum PIF capacity did not exceed 150 tons until 1977 and, as recently as 1981 only one project had been completed with a design capacity of 200 tons. In contrast, the minimum design capacity on PIF projects since 1983 has been about 150 tons.

While the trend has been one of increasing capacity, there has been very little change in the average driven lengths of PIFs which range from about 18 to 25 feet. The average driven length only exceeded 30 ft. on four projects. Prior to 1981, the average driven length was about 20 ft. compared to 23.5 ft. for projects completed since 1981 reflecting, in part, the trend towards higher design capacities over the last few years.

Except in a few cases where partially weathered rock occurred at shallow depths, all the PIFs are founded in the intermediate zone of residual soil as defined by Sowers (5). The PIFs were installed by either the bottom-driving or top-driving method as described later. Although PIFs have frequently been installed in groups, no predrilling was carried out before early 1984. Static load tests to twice the design load are available for about one-third of the projects.

Comparatively little has been published on the behavior of pressure injected footings in the Piedmont residual soils and weathered rocks, however, and the time is opportune for making a survey of the data.

It is the purpose of this paper to describe some recent findings from
1. Vice President, Engineering, Franki Foundation Company, Boston, MA
2. Manager-Geotechnical Services, Woodward-Clyde Consultants, Rockville, MD
3. Principal, Schnabel Engineering Associates, Bethesda, MD

two sites regarding pile capacity and the effects of installing nearby piles on the performance of previously completed piles. The changes in installation techniques that have recently been adopted in order to minimize pile heave are described and recommendations regarding modifications to the traditional pile test program are made.

TYPICAL SUBSURFACE CONDITIONS

The characteristic feature of subsurface conditions in the Piedmont region is the extreme variability in the depth of weathering and the thickness of the blanket of residual soils. It is not uncommon to find differences of up to 30 ft. in the elevation of partially weathered rock on a single site. The residual soil profile can be considered to consist of three zones:

1. The Upper Zone which is a crust of stiff sandy clay usually less than 10 ft. thick.
2. The Intermediate Zone, which consists of loose to dense micaceous sandy silt and silty sand, extends to depths of as much as 75 ft.
3. The Partially Weathered Zone, which represents the transition to unweathered rock, is normally identified as material where the standard penetration (SPT) resistance is more than 60 blows per foot.

It is the Intermediate Zone which is of primary interest in the design and installation of pressure injected footings. In this material, called residual soil, decomposition has advanced to such a stage that little visual evidence of the parent rock structure remains. However, towards the lower boundary of this stratum, foliation and jointing planes are more visible. The residual soil grades into a transitional material between soil and weathered rock. The concept of transitional material is that the changes between strata are in actuality very gradual and variable. The density of the soil generally increases with depth, although there are significant variations both laterally and with depth. Characteristics of the transitional materials include foliation and jointing planes and the presence of hard pockets or bands of rock-like material and occasional quartz boulders. The transitional material grades into weathered rock with depth, although the boundary between these materials is poorly defined. According to Sowers (5) the soil minerals are mainly quartz, clay minerals, partially weathered feldspars and mica. The mica is more resistant to weathering than feldspar and hence the Intermediate Zone materials often contain large, and variable, quantities of mica.

The range in grain size curves is illustrated in Fig. 1 for the Intermediate Zone soils as defined by Sowers (5) together with some typical curves from sites where PIFs have been installed. These materials are usually classified as ML in terms of the Unified Soil Classification System and the most common descriptions are micaceous sandy silt and micaceous silty sand.

Routine site investigation practice on Piedmont profile sites consists of the sinking of borings with SPT sampling at about 5 ft. intervals

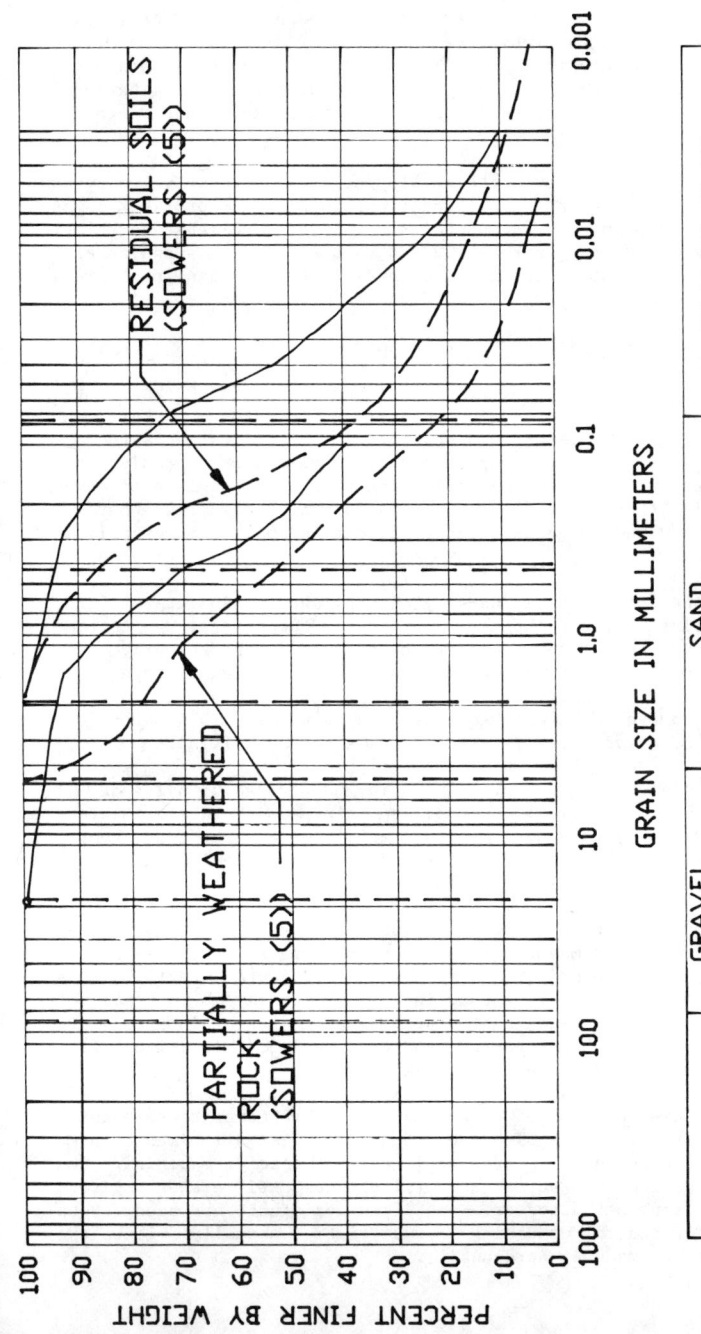

FIG. 1 TYPICAL GRAIN SIZE CURVES

usually until partially weathered rock is encountered. Occasionally, use is made of in situ pressuremeter tests for the evaluation of compressibility characteristics. Only a limited amount of laboratory testing (usually grain size analysis, index property determinations) is undertaken mainly because of the highly variable nature of the materials.

At present, the use of strength envelopes, such as those given by Sowers (5) in the design of deep foundations in Piedmont profiles is severely limited primarily because of the lack of a reliable correlation with standard penetration test (SPT) N values. For this and other reasons deep foundation design, and particularly the design of pressure injected footings, is largely empirical in nature.

INSTALLATION OF PIF IN PIEDMONT PROFILES

Practically all of the 25000 pressure injected footings founded in Piedmont materials are of the uncased type, i.e. having a shaft consisting of compacted zero-slump concrete. In most instances the shafts are unreinforced, except where it is necessary to resist uplift and/or lateral loads. The installation procedure consists of advancing the drive tube, making the base and forming the shaft. Where piles are required in groups, the minimum pile spacing is usually 4.5 feet.

Advancing the drive tube is usually accomplished by an internal drop hammer falling on a dense plug of zero-slump concrete at the bottom of the drive tube. After reaching the desired depth, the drive tube is withdrawn slightly to allow the plug to be expelled by repeated blows of the drop hammer. Sufficient plug material is left in the drive tube to avoid the entry of soil and/or water. This method of advancing the drive tube is referred to as bottom driving.

Drive tube diameters are typically 16 in. OD for medium piles and 21 in. OD for standard piles; uncased zero-slump concrete shafts are conservatively assigned nominal diameters 1 in. greater than the drive tube. Drop hammer weights range from 2.5 tons for medium piles to 3.5 tons for standard piles.

Depending on the strength and thickness of the strata overlying the bearing material, it may be necessary to utilize the top-driving technique to advance the drive-tube in a more efficient manner. In this case the bottom of the drive tube is closed by means of an expendable steel boot. Upon reaching the desired depth, zero-slump concrete is introduced into the drive tube and the steel boot is driven off by the internal drop hammer.

Making the base consists of expelling zero-slump concrete through the bottom of the drive tube with the drop hammer; at all times maintaining sufficient seal to prevent ingress of foreign material. Concrete is expelled in a series of small batches until either a given volume has been introduced or a specified resistance has been encountered. Field control of basing characteristics normally follows the recommendations made by Nordlund(4).

Forming the shaft of an uncased pile involves withdrawing the drive tube in a series of short steps while expelling concrete from the bottom of the drive tube. The internal drop hammer is used to compact the freshly placed concrete to produce a shaft of compacted zero-slump concrete. Shaft construction is continued in this fashion until the shaft is completed to the required cut-off elevation or to the ground level.

In some situations, it may be necessary and/or desirable to make use of a different type of shaft. Cased shafts, employing corrugated metal shells, have been used on a few projects in Piedmont profiles. The cased shaft is formed by seating the shell on the pile base prior to withdrawing the drive tube; the shell is subsequently filled with normal slump concrete.

The wet shaft technique has also been used in Piedmont materials. After making the base, the drive tube is filled with high slump (± 8 inches) concrete, leaving a shaft of wet concrete as the drive tube is slowly extracted.

BEARING CAPACITY OF PRESSURE INJECTED FOOTINGS

Early applications of pressure injected footings in Piedmont soils were based almost entirely on experience gained from working in naturally deposited sands, although it was recognized that design capacities should be somewhat less to account for the lower strength of the fine grained, micaceous residual materials.

With the increase in design loads over the past few years, it is now essential that some type of bearing capacity analysis be undertaken to provide more reliable estimates of founding depths, base sizes and safe working loads. Of the more than 30 load tests available from previous projects in the Piedmont region, only a few have been loaded to failure, usually after satisfactory completion of the preconstruction load test. These and other load test results have been analyzed in an attempt to formulate a simple, empirical design procedure for PIFs.

While it is recognized that effective stress type bearing capacity analysis may be more appropriate for the residual materials, which do not behave as saturated cohesive soils during undrained shear in a triaxial test, only a quasi $\emptyset = 0$ analysis has been attempted at this stage, primarily for the following reasons:

1. There are no correlations between SPT N values and relative density, void ratio and angle of shearing resistance of the residual soils, similar to those for naturally sedimented sands. Therefore, any design method must, of necessity, incorporate the raw N values directly into the bearing capacity analysis.

2. Experience indicates that on many sites, the Piedmont residual soils behave very much like saturated, insensitive clays. The most common characteristics are the occurrence of ground

heave (as well as heave of piles already installed) when PIFs are installed in large or closely spaced groups and erratic basing characteristics where, for example, the final PIF in a group requires a much larger volume of base concrete than the first PIF indicating, in part, that very little densification (or volume change) has taken place.

3. Except for a few cases, there is inadequate, reliable information on ground water level.

For uncased pressure injected footings, the ultimate bearing capacity, Q_u of a single isolated pile is the sum of the end bearing capacity, Q_e and the shaft friction capacity, Q_s. The end bearing capacity is calculated from:

$$Q_e = C_{ub} N_c A_b \dots\dots\dots\dots(1)$$

where the C_{ub} is the strength of the soil at the level of the PIF base, A_b is the area of the PIF base and N_c is a bearing capacity factor having a value of 9 for deep foundations in saturated, insensitive cohesive soils. The shaft bearing capacity is derived from:

$$Q_s = A_s C_a \dots\dots\dots\dots\dots(2)$$

where A_s is the area of the shaft in contact with the soil and C_a is the average adhesion between the compacted concrete shaft and the soil.

Two load tests were conducted in residual soil at a site in Fairfax, VA in order to determine, separately the end bearing and shaft friction loads. The results of these tests are presented in Fig. 2. Test PIF 1 had a cased shaft to ensure that the full test load would be supported at the base of the PIF. A total volume of 25 cu ft of zero-slump concrete was used to form the base, which is equivalent to a base diameter of 3.37 ft assuming a spherical base and a compaction factor of 0.8 on the bulk volume of base concrete. Test PIF 2 was constructed without a base in order to minimize end bearing resistance; the shaft consisted of compacted zero-slump concrete. Test PIFs 1 and 2 were loaded to failure at 225 tons and 200 tons respectively. The average standard penetration resistance along the shaft was N_s = 14 and at the base N = 23. Substituting the measured capacities in Eqs (1) and (2) it can be shown that the strength of the soil beneath the base is 2.8 tsf, or C_{ub} = 0.12 N (tsf), and the average adhesion along the shaft is 1.24 tsf, or C_a = 0.09 N_s (tsf).

The results of a further five load tests on cased and uncased PIFs at sites throughout the Piedmont region are summarized in Table 1. The values of ultimate capacity predicted using Eqs (1) and (2), using the values of C_{ub} and C_a derived from test PIFs 1 and 2, are seen to agree well with the measured capacities.

The quasi ∅ = 0 method of analysis has also been used to calculate the ultimate capacities of a number of PIFs that were not loaded to failure. For these PIFs the failure load was predicted from the available load settlement data using the stability plot technique suggested by Chin and Vail (1). Assuming that failure occurs at a settlement of 1 inch, this method gives about the same failure load as

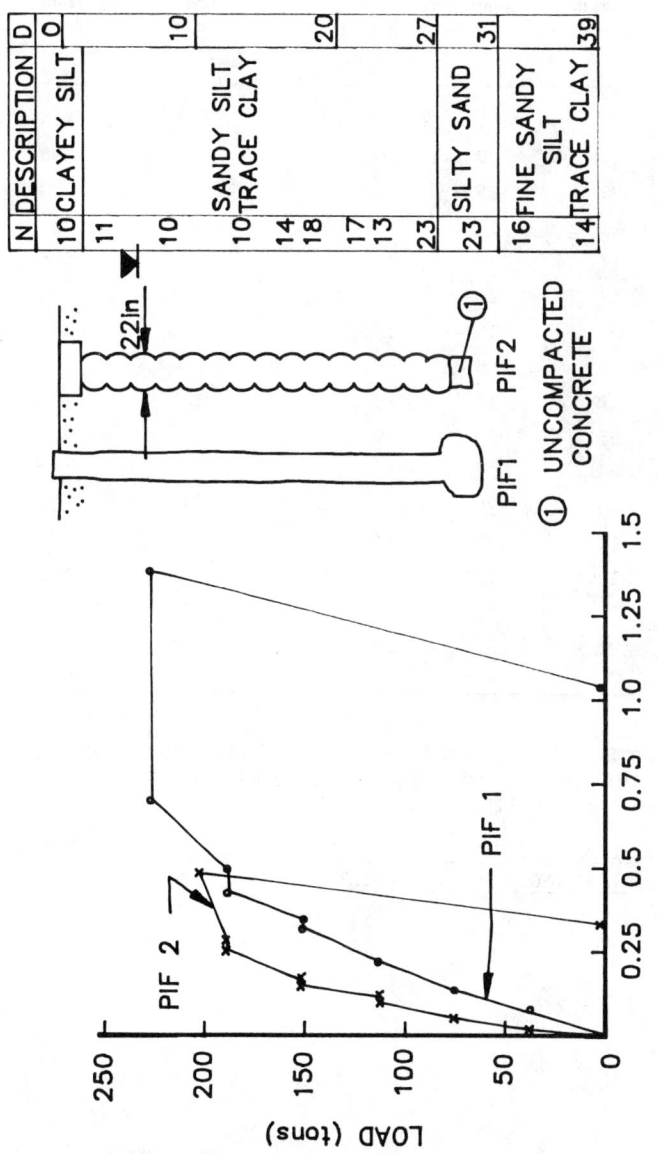

FIG. 2 LOAD SETTLEMENT CURVES FOR PIFS 1 AND 2

Test PIF	Driven Length ft.	Vol of Base Concrete, cu ft	N S	N	Measured Failure Load tons	Predicted Failure Load tons
3[a]	32	5	--	48	290	235
4	21	20	16	27	425	398
5	30	25	10	13	325	280
6	21	15	22	14	300	335
7	17	15	17	14	225	245
8[b]	17	10	9	21	182	171
9	21	15	13	19	334	272
10[a]	63	5	--	100	364	329
11	27	15	13	14	304	277
12	25	20	20	25	414	466
13	28	25	16	27	451	492
14	40	35	8	10	399	286
15	16	20	14	19	389	274
16	21.5	25	26	20	548	482
17[c]	34	15	29	23	509	668
18[d]	21	25	24	30	372	550

[a]Cased shaft; [b]Medium uncased shaft; [c]Predrilled to 31 ft; [d]predrilled to 16 ft. Measured failure loads for PIFs 8-18 estimated from stability plot analysis of load test data.

TABLE 1 SUMMARY OF MEASURED AND PREDICTED FAILURE LOADS OF PIFs IN PIEDMONT RESIDUAL SOILS

the van der Veen construction used by Nordlund (4).

The ultimate loads of fifteen pressure injected footings have been calculated using Eqs (1) and (2), where $C_{ub} = 0.12 \, N$ and $C_a = 0.09 \, N_s$, and are compared with the observed values in Fig. 3; the data from Table 1 are also included. In all but two cases the calculated values appear to be sufficiently close for the empirical method to be useful in practice for preliminary capacity evaluations for PIFs when considering several deep foundation options. However, in the case of two PIFs which were predrilled over most of the shaft length, the calculated capacities exceed the deduced failure loads by 35% and about 60%. The effects of predrilling on the capacity of uncased pressure injected footings will be discussed in detail later.

For routine PIF capacity evaluations it is considered that little accuracy is sacrificed if $C_{ub} = 0.1N$ and $C_a = 0.1 N_s$ are used in Eqs (1) and (2).

FIELD EXPERIENCES WITH PIF INSTALLATION

Previous Practice

Prior to 1984 the execution of PIF work was comparatively routine due to well established procedures based on wide experience with the pile throughout the Piedmont region. Most of the PIFs installed were of the uncased (and usually unreinforced) compacted concrete shaft type. The PIFs were frequently used in groups and installed using either the bottom-driving or top-driving method without any predrilling of pile locations. A load test on a single isolated pile was usually undertaken for most large projects. Control of pile driving operations was done by monitoring basing characteristics to ensure that base formation continued until a specified minimum blow-count had been achieved, generally in accordance with the pile driving formula for PIFs developed by Nordlund (4).

Recent Case Studies

In 1984 a PIF foundation was installed on two adjacent sites located in northern Virginia. Site A is located down slope of Site B, with the difference in elevation being approximately 20 feet. A statistical analysis of the SPT blow counts for the two sites is presented in Fig. 4, where it is evident that despite the 20 feet difference in elevation, both the degree and depth of weathering are about the same on both sites.

Pile Installation and Load Tests on Site A

In early 1984 production pile driving on a site (referred to subsequently as Site A) in northern Virginia was just under way when heave of the ground surface was observed. It was decided to check the top of the first PIF driven in both a 4-pile and a 5-pile group. The results, which are shown in Fig. 5a, confirmed that pile heave was occurring. Because there was no reinforcing in the compacted concrete shafts, concern was corrently expressed regarding the integrity of

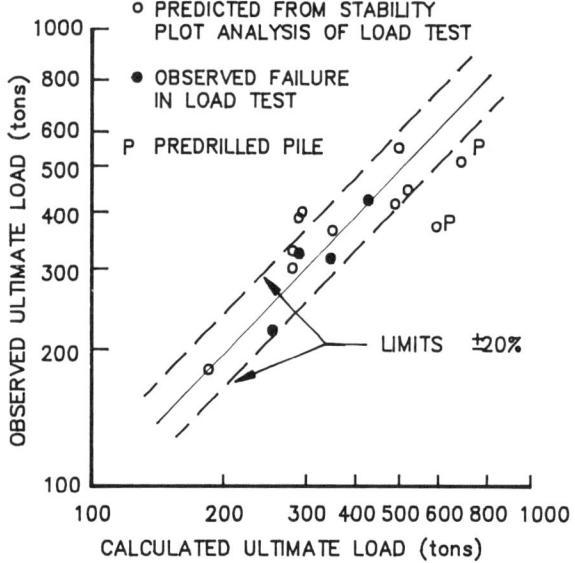

FIG. 3 COMPARISON OF OBSERVED AND CALCULATED FAILURE LOADS

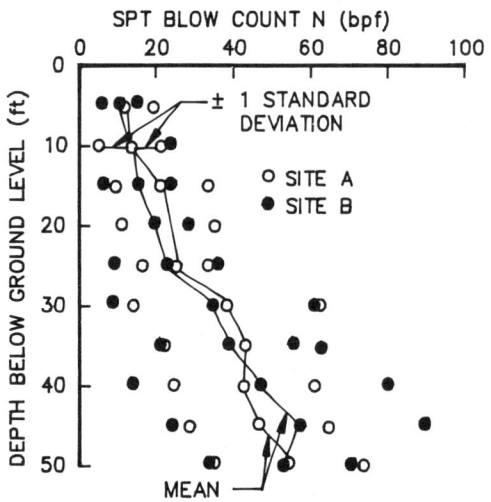

FIG. 4 STATISTICAL SUMMARY OF N VALUES FOR SITES A AND B

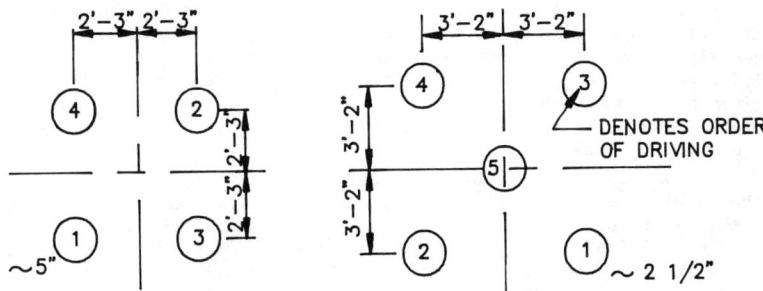

FIG. 5a PILE HEAVE WITHOUT PREDRILLING

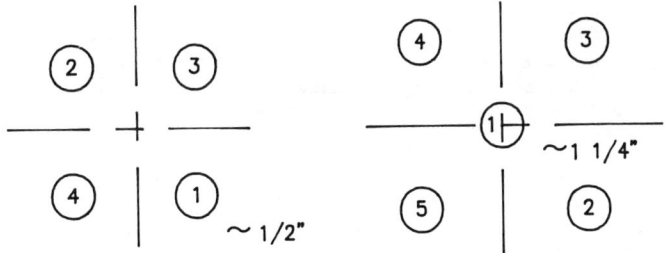

FIG. 5b PILE HEAVE WITH PREDRILLING

the shafts of heaved piles. For this reason a decision was made on site to predrill all remaining production piles to within 2 ft. of tip elevation. As indicated in Fig. 5b, this was very effective in reducing pile heave; the heave values for the 5-pile groups are not directly comparable because of the different driving sequences adopted.

The development at Site A required an excavation up to 20 ft. below original ground level. The buildings were designed to be supported on 200 ton capacity uncased, unreinforced compacted shaft type PIFs. A test PIF was installed from excavation subgrade to a depth of about 21.5 ft (El 422.5) using the bottom driving technique; the driving record and load settlement curve are shown in Fig. 6. A total of 25 cu ft of zero-slump concrete was used to form the base. The PIF satisfactorily supported the required test load of 400 tons and was subsequently loaded to 480 tons (the capacity of the test frame) without inducing failure. The stability plot method, referred to previously, gives a value of 550 tons for the failure load compared with 506 tons predicted by the empirical design method using the SPT N values from a nearby boring.

Pile heave was observed during installation of production piles even though all locations were predrilled to depths of about 15 feet, to within 2 ft. of tip elevation. The measured heave amounted to a maximum of 1.9 in. for the first pile installed in a group of 5 piles at 4-1/2 ft. on center. Two production PIFs were selected for static load testing; in each case the PIF selected was the first installed in a 5-pile group. A supplementary boring was put down within 5 ft. of each pile.

Fig. 7 illustrates the range in tip elevation for the production piles and the SPT N values in the residual soils at Site A. The load-settlement behavior of the two heaved piles is shown in Fig. 8 where it can be seen that they settled between 1 1/2 in and 2 in at the 200 ton design load. The ultimate bearing capacity of the piles is about 200 tons corresponding to the beginning of the substantially straight portions of the load-settlement curves.

Both PIFs were founded slightly higher than the original test PIF but appear to bear on somewhat more competent residual material. A total of 15 cu ft of zero-slump concrete was used in making the base of PIF #A1 and 20 cu ft was used for PIF #A2. In an attempt to identify the cause of failure PIF #A1 was exhumed. It was found that the unreinforced shaft was undamaged and there was no separation of the shaft from the base. The concrete was well compacted and of sufficient strength. The base of the PIF was not spherical, as is normally the case, but tapered varying in diameter from about 29 in at the top to 26 in at the bottom; the base was about 33 in high. Assuming the base is a sphere, 15 cu ft of concrete should produce a base of at least 34 in. diameter.

It was reasoned that since the piles were predrilled the measured heave was the result of soil displacement caused by forming the bases of piles. As pointed out by Klohn (3), such displacement may result in appreciable heave but since this heave is caused by a general upward movement of the bearing stratum soils it is unlikely to result

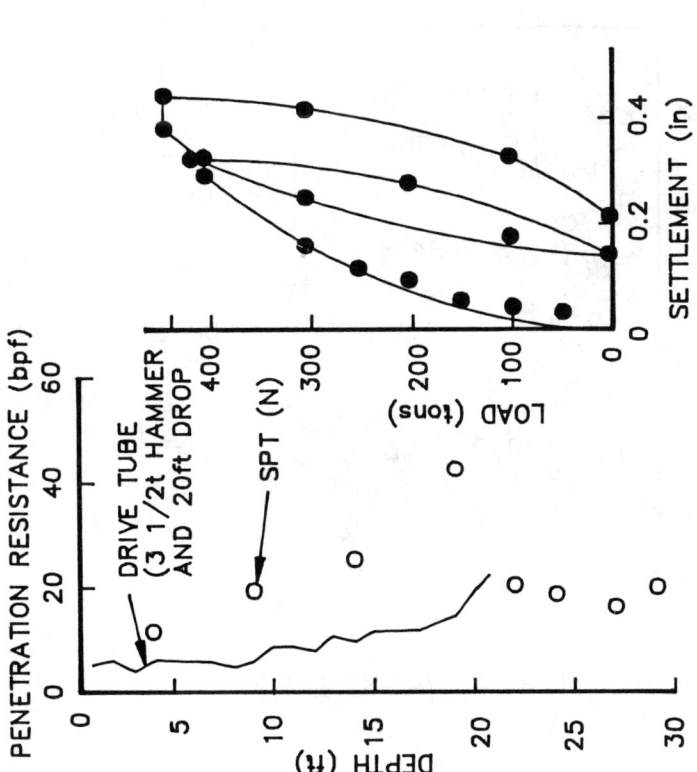

FIG. 6 PRECONSTRUCTION TEST PILE ON SITE A

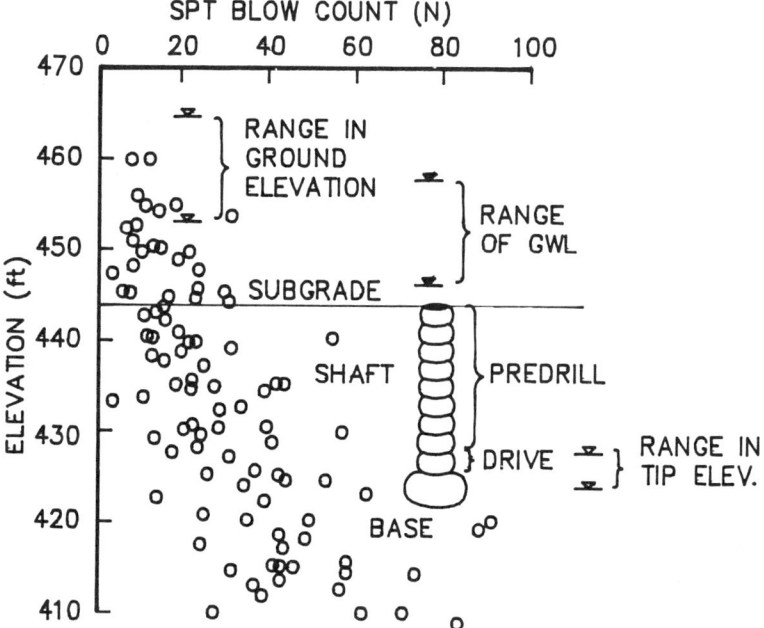

FIG. 7 PIF INSTALLATION ON SITE A

in a decrease in the resistance that the pile can develop within the bearing stratum.

The reasons for the poor performances of PIFs #A1 and #A2 are not believed to be directly related to pile heave, but rather to a reduction in pile bearing capacity due to the following factors:

(i) Predrilling would have the effect of lowering shaft friction resistance because of reduced lateral stresses or the creation of a "smear" of remolded soil on the walls of the hole.

(ii) The tolerance on the depth of predrilling (to within 2 ft of tip elevation) was insufficient to ensure that the pile base would not be formed in loose drilling spoil or disturbed material at the bottom of the hole.

(iii) Even if the drive tube penetrated 2 ft below the bottom of the predrilled hole, there would be insufficient skin friction between the soil and the drive tube to prevent the tube being lifted as the base concrete is forced out. As was confirmed, this would create a smaller, non-spherical base, thereby reducing end bearing resistance.

(iv) The time between predrilling and forming the pile is likely to be important, particularly where predrilling extends below the water table. It is essential that this be kept to a practical minimum to avoid water ingress and softening of the soil below the bottom of the hole. In practice, this can be achieved with predrilling equipment mounted directly on the piling rig.

(v) The shaft lengths of the two production PIFs tested were less than that of the original test PIF which was completed to ground level to facilitate load testing.

The effect of a smaller base and shorter shaft length on the capacity of PIF #A1 has been investigated using Eqs (1) and (2) and the N values shown in Fig. 8. Ignoring any reduction in shaft friction resistance due to predrilling, then for a shaft length of 15 ft and a base diameter of 29 in, the empirical design method gives an ultimate capacity of 300 tons. A similar analysis for PIF #A2, assuming a base diameter 5 in smaller than for a sphere of the same volume, yields a failure load of 337 tons.

Therefore even without any allowance for reduced shaft friction, it is unlikely that the production PIFs could have supported the 400 test load.

52　FOUNDATIONS AND EXCAVATIONS

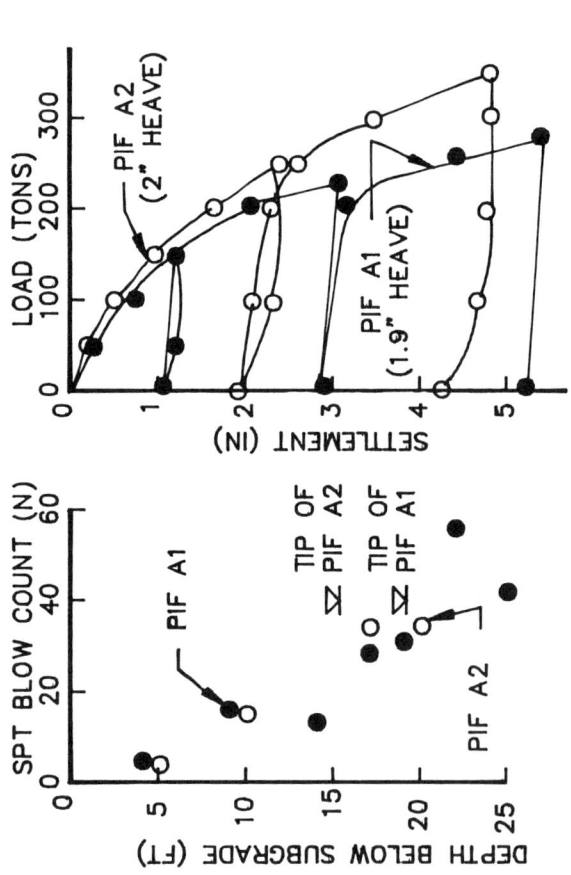

FIG. 8 LOAD TESTS ON HEAVED PIFS ON SITE A

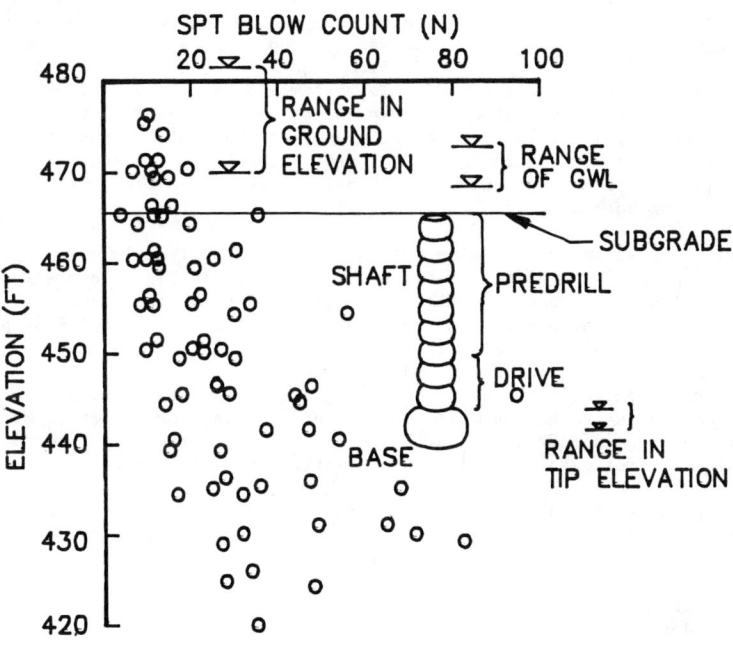

FIG. 9 PIF INSTALLATION ON SITE B

Because of time constraints and uncertainty about pile capacity due to variations in shaft length, depth of predrilling and base size and shape, the owner decided to abandon the piles already installed and the building was completed on a mat foundation.

Pile Installation and Load Tests on Site B

As a result of observed heave on a project completed in 1982, which was handled contracturally, there was a re-evaluation of the installation methods, pile testing program and the nature and scope of inspection services planned by the geotechnical consultants for the next PIF project at Site B. First, it was recognized that predrilling would probably not eliminate pile heave and that this issue should be addressed in both the test pile program and throughout pile driving operations. For this reason it was decided to load test the first pile installed in a group of 3 piles (and hence subjected to heave) rather than a single isolated pile that would not be heaved. Second, in order to ensure that a properly shaped base would be formed in undisturbed soil, it was decided to advance the drive tube not less than 5 or 6 ft. below the bottom of the predrilled hole. In addition, there was a considerable amount of interaction between the professional teams for both projects.

Fig. 9 presents the SPT blow counts in the residual soils on Site B; a comparison with similar data for Site A given in Fig. 7 shows that the piles are founded in soils of roughly the same consistency at both sites. The preconstruction test PIF configuration is illustrated in Fig. 10a which also indicates the installation sequence. Fig. 10b shows that test PIF 102 heaved 1 in. due to installation of the other two piles in the group. Each of these piles resulted in 0.5 in. of heave of the test pile with about 0.25 in. due to bottom driving the tube 5 1/2 ft. below the bottom of the 16 ft. deep predrilled holes. Forming the base produced a further 0.25 in. of heave.

The results of a 300 ton (twice design load) static load test on Test PIF 102 are presented in Fig. 11 where it is evident that the performance of the pile was not adversely affected by heave.

Full-time monitoring of production PIF installation was carried out in order to document predrilling and driving operations, pile heave and basing characteristics. Out of a total of 224 PIFs, only 23 heaved more than 1 in. and only 4 PIFs heaved in excess of 1.5 in; the maximum heave was just over 2 in. A post-construction load test was carried out on a PIF that had heaved 1.8 in; the results are included in Fig. 11. Again, there is no evidence that heave had any effect on performance although the PIF settled a little more than the original test pile. This is probably accounted for by the smaller volume of concrete used in forming the base and the shorter shaft length of the production PIF.

The heave measurements for Site B are summarized in Fig. 12 where it can be seen that while there is considerable scatter, probably due to variations in the number of piles driven within the zone of influence around the observation pile, there is a trend of increasing heave as the driven depth is reduced. This is to be expected since at

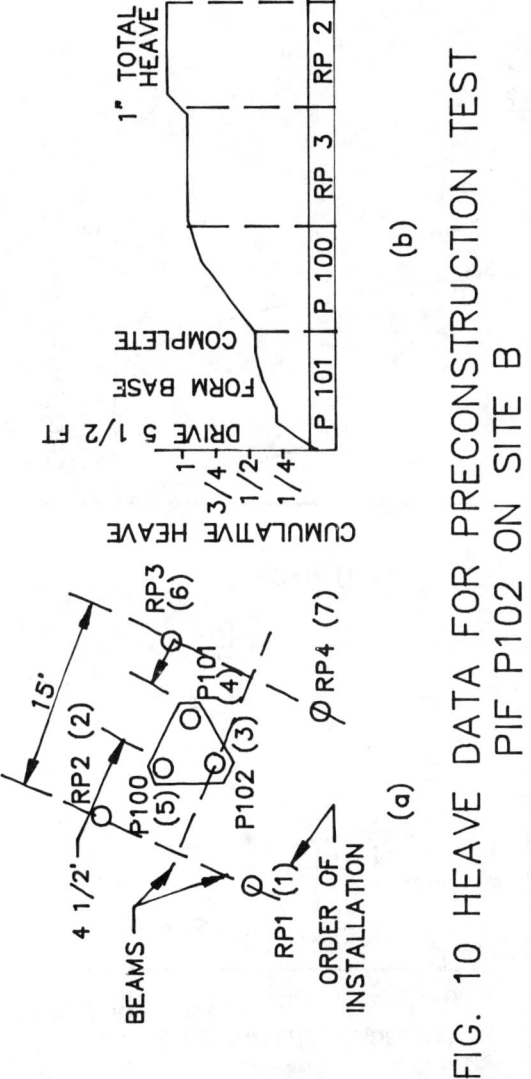

FIG. 10 HEAVE DATA FOR PRECONSTRUCTION TEST PIF P102 ON SITE B

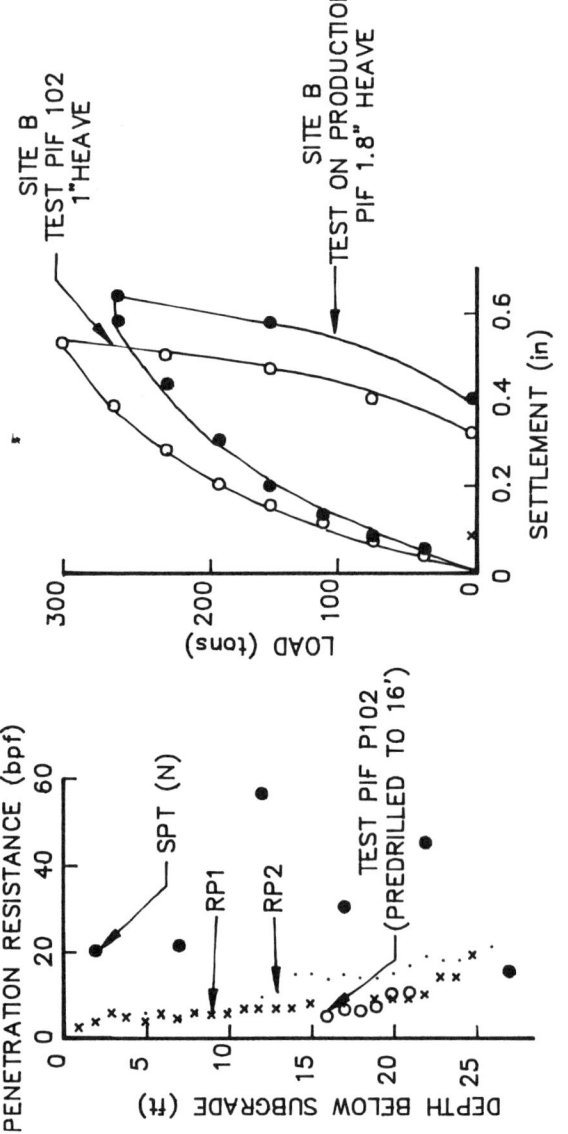

FIG. 11 LOAD TESTS ON HEAVED PIFS ON SITE B

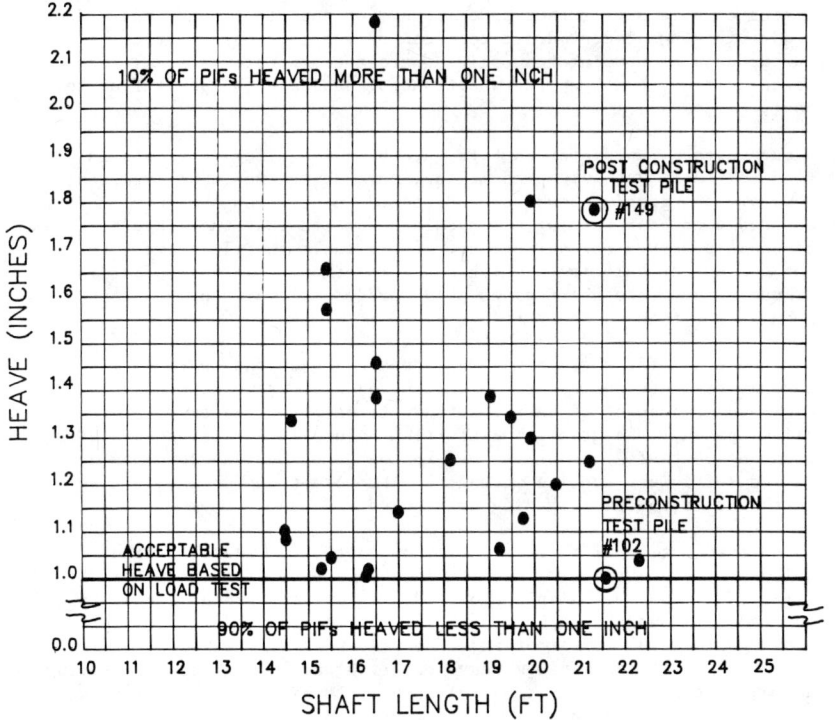

FIG. 12 SUMMARY OF PIF HEAVE ON SITE B

shallow depths soil displacement due to driving is predominantly upwards and a short pile would have less resistance to heave than a pile driven below the zone of upward soil displacement. In addition, nearby piles formed below the base of a short pile would tend to produce greater heave.

GENERAL DISCUSSION AND CONCLUSIONS

Although pressure injected footings have been used in Piedmont profiles for almost 20 years, the problem of pile heave has only come to light in the past few years. The heave that occurs due to forming an adjacent expanded compacted concrete base appears to have no direct effect on pile performance under load. On the other hand, heave due to driving an adjacent drive tube can be very serious, particularly in the case of unreinforced piles where the concrete is still fresh, because it may break up the shaft and/or separate the shaft from the base.

Since uncased, unreinforced PIFs have been installed in groups in Piedmont soils without predrilling, some damage would be expected; however, the problem has gone undetected on previous projects. Having sounded a note of warning, it should be stated that there are no cases of unsatisfactory PIF performance attributable to pile heave. Part of the reason for this is the high shaft friction resistance of driven, compacted concrete shaft piles which are capable of supporting as much as 200 tons in shaft friction.

The potential for shaft damage and separation of the shaft from the base is greatest during driving of an adjacent drive tube because the upward displacement of the soil tends to pull the shafts of nearby piles upward. The greatest distortion occurs in unreinforced shafts when adjacent PIFs are driven before the compacted concrete has acquired sufficient tensile strength. The potential for damage also increases as the pile spacing is reduced.

As pointed out by Clark et al. (2), at depths in excess of about 8 drive tube diameters (i.e. 14 ft. in the case of the 21 in OD standard drive tube) the soil displacements due to driving are no longer upward, but radially outwards. Driving below this depth produces no significant upward displacement of adjacent pile shafts. This was confirmed at Site B where the maximum heave observed was 1/4 in. due to driving an adjacent pile (4 1/2 ft. away) about 6 ft below the bottom of a 15 ft. deep predrilled hole.

Since potentially damaging upward soil displacements occur only to a limited depth then, if the ultimate capacity of the shaft above this depth is less than the sum of the ultimate capacity of the shaft below this depth and the tensile strength of the shaft/base connection, the shaft will not separate from the base. Resistance against shaft uplift can be improved in two ways. First, increasing the spacing between adjacent piles driven in a given period of time,(e.g.12ft spacing for piles driven on the same day) would allow the zero-slump concrete to gain some tensile strength. Second, the shaft/base connection can be strengthened by means of a reinforcing cage securely anchored in the base. In practice, it is usually more cost effective to predrill pile locations.

The difference between the load settlement behavior of heaved piles on the two study sites is clearly illustrated in Fig. 13. As noted earlier, a comparison of SPT blow counts showed that subsurface conditions are practically identical for both sites. The information provided in Fig. 13 also shows that both piles heaved the same amount eliminating heave as a primary factor in determining pile performance under load. In addition, both piles were predrilled leaving the depth of driving below the bottom of the predrilled hole, the volume of base concrete and the final shaft length as the only major differences The pile exhumed at Site A revealed that 2 ft of penetration below the bottom of the predrilled hole was insufficient to ensure a properly shaped base of adequate size. At Site B all piles were based in undisturbed soil about 6 ft. below the bottom of the predrilled holes.

At Site B less emphasis was placed on achieving a specified blow count to expel each 5 cu ft of base concrete as the basis of pile acceptability. Instead it was required that each PIF base should have a minimum volume of 20 cu ft of zero-slump concrete (including the plug) before applying the pile driving formula for PIFs developed by Nordlund (4). For this reason pile bases on Site B are probably slightly larger than on Site A - possibly by as much as 10-12 in.

The major findings from the case histories described herein that have direct influence on the design, load testing and field inspection for PIFs installed in Piedmont profiles include the following:

1. Where PIFs are to be installed in groups, all pile locations should be predrilled at least 15 ft. to minimize potentially damaging upward soil displacements. This is considered a practical minimum; depending on soil conditions, pile spacing and capacity it may be necessary to predrill to greater depths. Every effort should be made to keep the time between predrilling and making the pile to a minimum.

2. The drive tube should be advanced at least 5 ft. below the bottom of the predrilled hole to ensure that a base of appropriate size and shape is formed in undisturbed material.

3. The depth of predrilling will also depend on the depth of pile cut off below working grade and should be chosen such that the final shaft length of the pile is not less than 15 ft.

4. Before applying any dynamic pile driving formula, the pile should be advanced to the minimum tip elevation determined from static analysis and a minimum of 15 cu ft of concrete should be expelled to form the base.

5. The preconstruction load test program should, ideally consist of installing all the piles in a typical sized group and conducting a static load test on the pile that

POST CONSTRUCTION PIF LOAD TEST

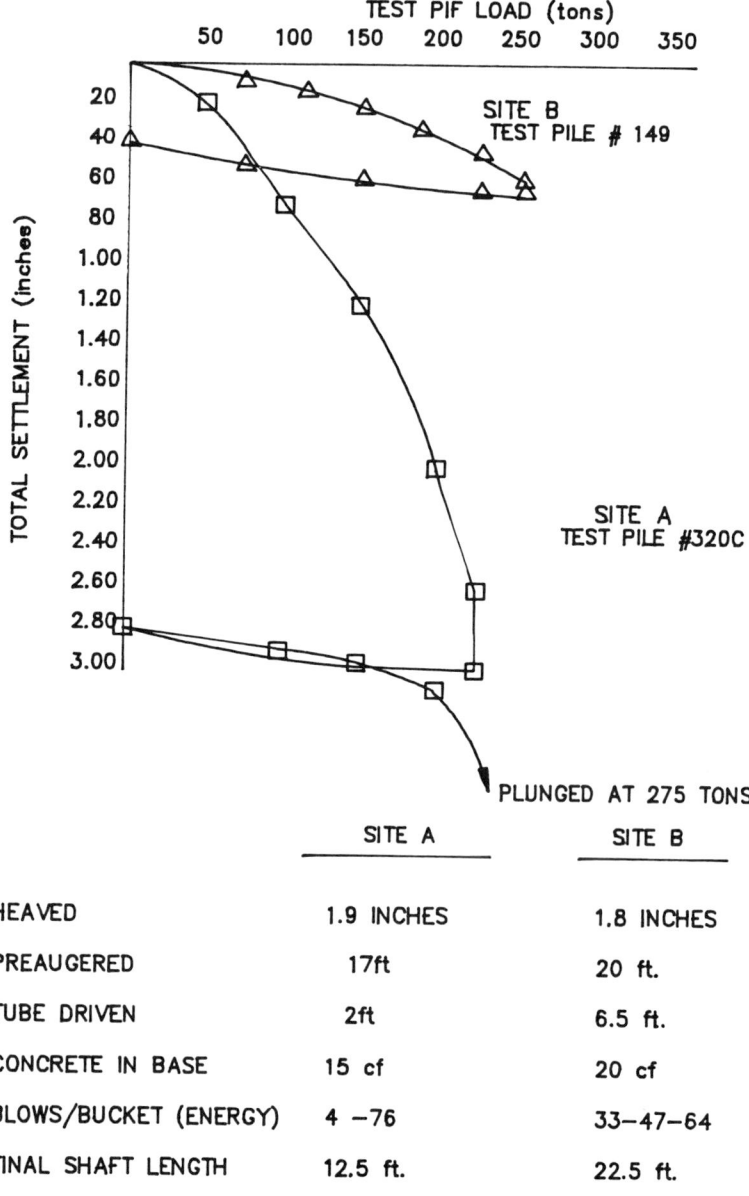

FIG. 13 COMPARISON OF HEAVED PIFS AT
SITE A AND SITE B

was installed first.

6. Full time on site monitoring of production PIF installation by qualified and experienced geotechnical personnel is strongly recommended so that installation procedures, founding depths and basing characteristics and pile heave can be properly documented.

APPENDIX 1 - REFERENCES

1. Chin F.K., and Vail, A.J., "Behavior of Piles in Alluvium" Proc, 8th Int. Conf. on Soil Mechanics and Found. Engng., Moscow, 1973, Vol. 2.1, pp. 47-52.

2. Clark, J.I., Harris, M.C., and Townsend, D.L., "Heave of Compacted Expanded Base Concrete Piles", Proc. 10th Int. Conf. on Soil Mechanics and Found. Engng., Stockholm, 1981, Vol. 2.1, pp. 667-672.

3. Klohn, E.J., "Pile Heave and Redriving", Proc. ASCE, Vol. 87 No. SM4, Aug., 1961.

4. Nordlund, R.L., "Dynamic Formula for Pressure Injected Footings," Journal of Geotechnical Engineering, ASCE, Vol. 108, No. GT3, Mar. 1982, pp. 419-437.

5. Sowers, G.F., "Engineering Properties of Residual Soils Derived from Ingeous and Metamorphic Rocks," Proc. 2nd Panamerican Conf. on Soil Mechanics & Found. Engng., Sao Paulo, 1963; Vol. 1, pp. 39-61.

DESIGN OF DRILLED PIERS IN THE ATLANTIC PIEDMONT

William S. Gardner*, M.ASCE

ABSTRACT

The philosophy and methods of drilled pier foundation design in the Atlantic Piedmont are described. Consideration is also given to the unique characteristics of decomposed metamorphic rocks of this region and to a design approach which accommodates unanticipated variation of the weathering profile.

INTRODUCTION

A large area of the eastern United States extending from Georgia northward to New York State is characterized by outcrops of Precambrian metamorphic rocks. As shown by Figure 1, this area is generally associated with the Piedmont Physiographic province of the U.S. The foundation conditions unique to this region involve a weathering profile whose prime characteristic is unpredictable variability in both the thickness and quality of the weathered materials.

Starting with the major economic developments within the Atlantic Piedmont following World War II, the application of drilled pier** foundations for major structures has increased significantly. Today, drilled piers are used to support the majority of the major structures being constructed in this region.

The approach to drilled pier design varies considerably throughout the Piedmont, and often, with good reason. However, in some instances local precedence has inhibited the application of more cost-effective designs and constructions, which are possible with modern pier drilling equipment and tools.

The primary intent of this paper is to describe a philosophy and approach to drilled pier design which has proven successful for a variety of projects located throughout the weathered metamorphic rocks of the eastern U.S. Consideration is also given to the characteristics of the weathering profile, the associated materials and their properties.

SUBSURFACE CONDITIONS

The principal rock types comprising the metamorphic rocks of the eastern U.S. are Precambrian mica schist/gneiss and the quartz

*Executive Vice President, Woodward-Clyde Consultants, 5120 Butler Pike, Plymouth Meeting, PA 19462.

**Also termed drilled caissons, drilled shafts and bored piles.

feldspar schist of the Inner Piedmont. These rocks have been subjected to extensive deformations and faulting and have been intruded by younger igneous plugs, dikes and sills which are usually more resistant than the host rock. These intrusives are often granite but include diabase, amphibolite, meta gabbro and other igneous rocks.

The Weathering Profile: Weathering processes have altered the parent rock, in some cases, to depths greater than 100 feet. Weathering has also been promoted along fault and fracture zones, sometimes to relatively great depths. In general, three weathering horizons can usually be identified. They are: (1) residual soil, representing advanced chemical alteration of the parent rock; (2) highly altered and leached soil-like material (saprolite) retaining some of the structure of the parent rock; and (3) decomposed rock, which is less altered but can usually be abraided to sand and silt-size particles. The underlying intact rock near the rock surface is usually fractured and weathered but increases in quality with depth. The interfaces of each horizon are not well-defined and their thicknesses may be highly variable.

Although, in general, there is a trend towards deeper weathering profile in the southern Piedmont, it is not unusual to encounter differences in the depth to intact rock over 50 feet in distances as little as 20 feet in the schists of Philadelphia, Wilmington and Baltimore. Examples of weathering profiles in Pennsylvania, Maryland and South Carolina are shown in Figure 2.

Residual Soil and Saprolite: Residual soils are the product of advanced chemical alteration of the parent rock and do not exhibit relic structure. They usually range from loose to medium dense micaceous silty sands (SM) to sandy clays (SC) and tend to become more granular with depth. The degree and depth of the alteration depends upon the climate, structure and facies of the parent rock. For example, a deeper and more plastic soil horizon would be expected in the southern Piedmont.

The saprolite zone immediately below the residual soil horizon zone may reflect intense leaching and have an unusually high void ratio (1). Such intense leaching is usually encountered in the southern part of the Atlantic Piedmont, but not always. Generally, the saprolite tends to become more granular and dense with depth, grading into decomposed rock.

Decomposed Rock: The dense to very dense friable horizon termed decomposed rock has a (relic) structure reflecting the schistosity and banding of the parent rock but has undergone chemical alteration of the rock minerals. There is a trend towards increasing penetration resistance (decreasing alteration) with depth, although there is sometimes little change in resistance until just above the intact rock surface. The decomposed rock horizon is of primary interest for foundation bearing, and relatively high loads can be supported, particularly in the lower part of the horizon.

Intact Rock: The schist/gneiss is usually defined as intact rock if RQD values (2) are greater than about 20 percent and

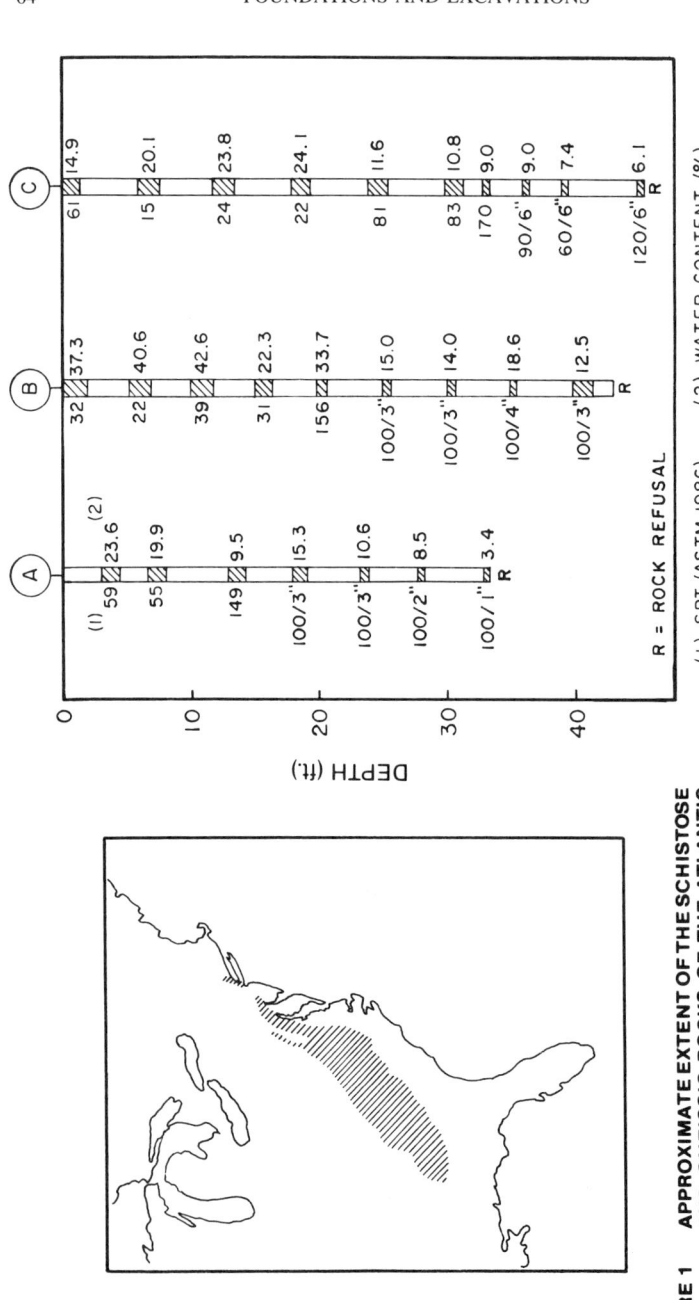

FIGURE 1 APPROXIMATE EXTENT OF THE SCHISTOSE AND GNEISSIC ROCKS OF THE ATLANTIC PIEDMONT

FIGURE 2 SAPROLITE AND DECOMPOSED ROCK PROFILES IN PA. (A), MD. (B) and S.C. (C)

the minerals of the rock core are not significantly altered by weathering. The quality of the rock, based on Rock Quality Designation (RQD) and hardness ratings (2) (3), can be quite variable but usually increases with depth. Average RQD's >50 percent usually represent very high capacity end-bearing provided there is no significant alteration of the intact rock minerals.

Groundwater: Depending on local conditions, groundwater may or may not be present within the depth of interest to drilled piers. Perched groundwater is, however, often encountered near, at, or just above the surface of the dense decomposed rock. This material usually has a relatively low permeability ($k < 10^{-5}$ cm/sec) and may serve to retard downward percolation. In addition to construction concerns, a perched water level is of significance in estimating the effective overburden stress through the decomposed rock horizon.

MATERIAL CHARACTERISTICS

The primary materials of interest to the support of relatively high loads by drilled piers are the dense to very dense decomposed rock horizon and the underlying bedrock. For decomposed rock, properties such as density, specific gravity, water content and the Standard Penetration Resistance (N_{SPT}) are useful to provide correlations with strength, compressibility and their variation. One-dimensional consolidation tests on undisturbed samples may also be of help in evaluating the current and maximum past vertical and horizonal stresses (state of stress).

Physical properties are also useful for the characterization of the intact rock. Additionally, it is appropriate to quantify the quality of the rock mass using classification systems which consider the frequency of rock mass discontinuities and to relate rock mass quality to strength and compressibility parameters. Alternatively, in situ tests provide a direct measurement of rock mass compressibility and strength.

Shear Strength of Decomposed Rock: Based on drained (CID) and undrained (CIU) triaxial compression tests on representative samples from a variety of locations, the effective peak friction angle (ϕ') of the dense ($60 > N_{SPT} < 100$) to very dense ($N_{SPT} > 100$) decomposed rock usually ranges between 29 and 32° and does not appear to be sensitive to void ratio. More limited data suggests that the friction angle at critical state (ϕ'_{cs}) is usually only slightly smaller than ϕ'.

The effective cohesion (c') of the decomposed rock generally increases with increasing density or water content, as well as penetration resistance. The range of c' values can be as little as 20 (medium dense to dense) occasionally exceeds 200 KPa (very dense). Figure 3 demonstrates a representative CIU test on very dense decomposed mica schist.

An example of the derivation of undrained shear strength from UU and PMT tests conducted in a very dense mica schist is shown by Figure 4. Dry density or saturated water content appear to be good

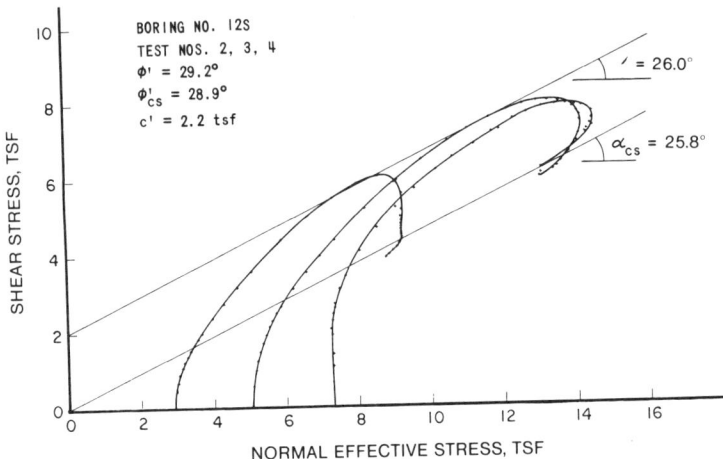

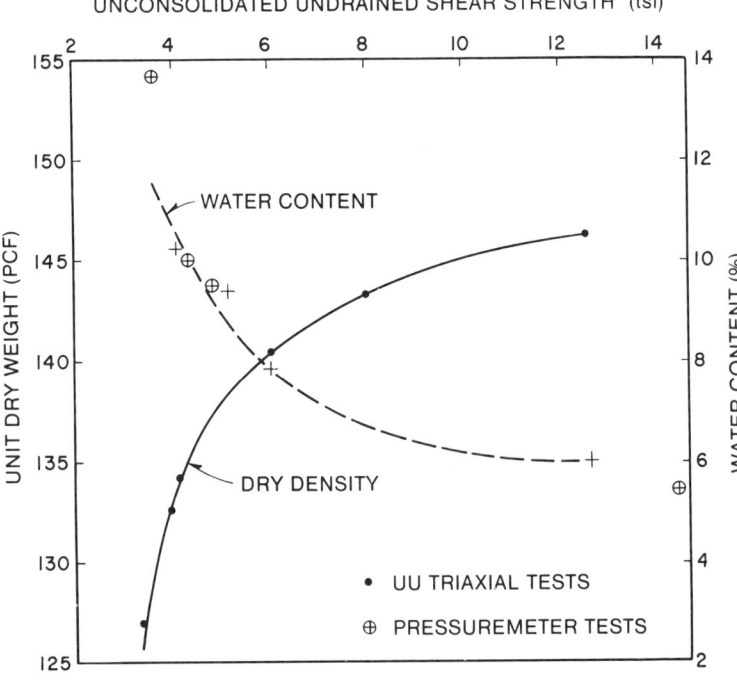

FIGURE 4 UNDRAINED SHEAR STRENGTH CORRELATIONS FOR VERY DENSE DECOMPOSED MICA SCHIST / GNEISS WITH N_{STP} = 100/6" TO 100/0"

correlators. The stress-strain curves of UU and $\overline{\text{CIU}}$ tests on very dense decomposed rock demonstrate the post-peak strength degradation common to dilative soils. Significant negative pore pressures are generated during undrained shear by the CIU test whereas CID tests on the same materials exhibit volume increase during drained shear.

Deformability of Decomposed Rock: The deformation (E_s) and shear moduli (G) of the decomposed rock have been derived from UU triaxial compression, one-dimensional consolidation and pressuremeter tests. Although E_s or G can be expressed in terms of drained or undrained moduli, there is usually not a significant difference for the dense to very dense decomposed rock. Figure 5 shows a comparison of the moduli derived from the three test methods as a ratio of the depth of the test (or test specimen) below the decomposed rock surface to the thickness of the horizon. The PMT and consolidation tests were interpreted from the reload data.

Deformability of Intact Rock: In situ tests relevant to the direct determination of the intact rock mass deformation modulus (E_{rm}) include pressuremeter, bore-hole jack and plate-bearing tests (4). The first two of these test types are designed for bore-hole testing and are usually the most cost effective. Seismic wave transmission measurements are also an indirect in situ test method to evaluate E_{rm} (5) (6).

The most frequently used method of rating the quality of a rock mass is Rock Quality Designation (RQD) proposed by Deere, et al, (6). E_{rm} has been related to RQD (%) as a function of the deformation modulus of intact rock core (7) as:

$$\frac{E_{rm}}{E_{50}} = 0.0231 \text{ RQD} - 1.32 = > 0.15 \qquad (1)$$

where E_{50} represents the tangent modulus of the stress-strain curve at one-half of the ultimate stress. Similar and more elaborate approaches using the Rock Mass Rating System are proposed by Bieniawski (8) and Chappell and Maurice (9) as:

$$E_{rm} = a \text{ RMR} - b \qquad (2)$$

where E_{rm} is given in units of MPa. For RMR values between 55 and 100, a = 2 and b = 100; for RMR <55, a = 0.16 and b = 3.

Strength of Intact Rock: Direct measurements of the shear strength of the rock mass are seldom made for drilled pier projects. However, small scale load tests have been conducted to measure the shearing resistance between a rock mass and a small diameter test shaft (10). Where it is appropriate to evaluate the ultimate end-bearing of a rock based pier (11), the strength of the rock mass can be estimated in terms of effective stress strength parameters c' and φ' for intact rock cores. ϕ'_{rm} can be estimated from φ' and the Rock Mass Rating by application of a reduction factor (β_ϕ) given as:

$$\beta_\phi = 0.005 \text{ RMR} + 0.45 \qquad (3)$$

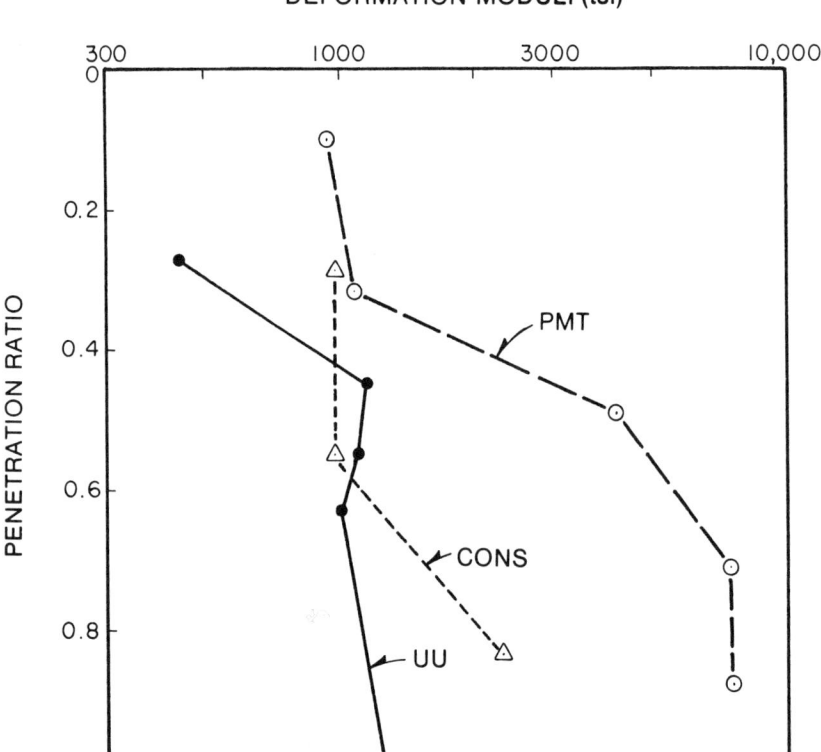

FIGURE 5   DEFORMATION MODULI VS TEST DEPTH / THICKNESS OF DECOMPOSED ROCK HORIZON FOR DIFFERENT TEST PROCEDURES

The c' parameter can be reduced by the same means used to estimate E_{rm}. Limited regional core test data suggests $\phi' = 48\text{-}55°$ for mica gneiss and $43\text{-}50°$ for mica schist. The inexpensive Uniaxial Compressive strength (q_u) and the Point Load Index (4) are useful for a variety of strength corrections, including estimation of c' by equation (4):

$$c' = \frac{q_u}{2\tan(45 + \phi'/2)} \tag{4}$$

The accuracy of this prediction depends on the quality of the ϕ' estimate.

DESIGN PHILOSOPHY

The philosophy of design developed from long-term experience with drilled pier foundations in the decomposed and intact micaceous schist/gneiss of the Atlantic Piedmont can be simply stated as:

(1) the primary design should be the one which utilizes the most cost-effective method of transferring the load from the shaft to the bearing materials together with an adequate and reliable safety margin,

(2) design and construction provisions must accommodate unpredictable variations in bearing materials and bearing levels, and

(3) the construction specifications and contract documents must contain provisions to facilitate field changes in the type of piers and the installation methods.

Each of these concepts are subsequently amplified.

Primary Design

To demonstrate primary design considerations, Figure 6 shows a hypothetical profile and three belled and three straight shafts carrying the same load. Option 1 is the shortest pier but the very large bell is subject to collapse in the residual soil. Underreaming of Option 3 in the weathered rock zone is at best very difficult and may require manual excavation. Option 6 requires intact rock drilling to form a rock socket and is usually competitive only where the design loads are quite large or there are no other suitable bearing materials. The remaining options are design candidates whose selection will depend on the rate of the installation, the amount of concrete and reinforcing steel required.

Except where the decomposed rock horizon is not sufficiently thick, straight and socketed shafts mobilizing significant skin friction, usually prove to be the most cost-effective solution. In this evaluation, it is imperative that consideration be given to the load-deformation compatibility of the pier and how this affects the shaft friction and end-bearing developed at working loads.

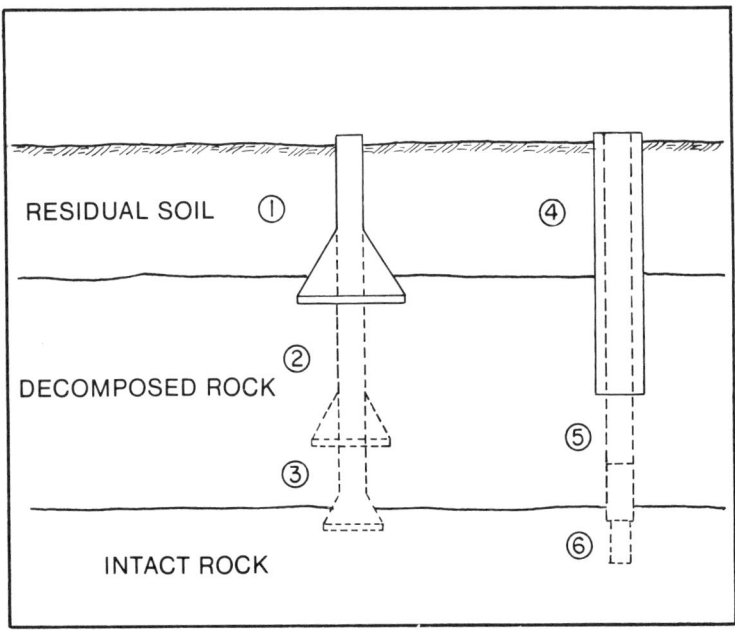

FIGURE 6   ALTERNATIVE DRILLED PIER TYPES IN PIEDMONT WEATHERING PROFILE

Accommodation of Variation

On some projects there are frequent, large variations in the thickness and quality of the decomposed rock horizon, although this frequency can sometimes be reasonably predicted by a thorough subsurface exploration. A construction which provides for more than one drilled pier type often results in large savings, if field changes can be made to accommodate different bearing conditions without delay. Consequently, design options should be incorporated in a manner to facilitate field application.

Specification and Contract Provisions

To implement design and construction options, it is necessary to incorporate appropriate provisions in the plans, specifications and contract documents. For example, the specifications should allow application of design options in the field at the discretion or the approval of the Engineer. The plans should provide the details of the design option and the contract documents should contain appropriate methods of payment for all such field changes. It is also desirable to permit the Contractor to bid on the design option he believes to be the most cost-effective.

DESIGN APPROACH

There are several approaches to the design of drilled piers. These include, the use of local precedence, load tests of prototype or small-scale piers, and analyses based on soil/rock properties and/or correlations. As load tests are usually cost prohibitive on most projects and there may be no precedent for design at a specific site, the focus here will be upon the latter method of analyses.

Limit State Analysis

The most common approach to static analysis for drilled pier design is to predict the ultimate pier load (P_{ult}) and apply a suitable global safety factor, usually between 2 and 2.5. The ultimate load is calculated by equation 5 as a function of the ultimate shearing resistance (τ_{su}) and end-bearing (q_{bu}) of the pier.

$$P_{ult} = \pi d_s \sum_{i=1}^{n} \Delta L (\tau_{su})_i + \frac{\pi}{4} d_b^2 q_{bu} \qquad (5)$$

In this equation, d_s and d_b are the shaft and base diameters, ΔL represents the incremental shaft length, and n the number of shaft increments used to represent the variation of τ_{su} along the shaft.

Notwithstanding the uncertainties in calculation of τ_{su} and q_{bu}, the primary deficiency in the ultimate load method is the uncertainty in the evaluation of the amount of load carried by the shaft and the base under service load conditions. A rational design of a drilled pier or pile cannot be developed without consideration of load-deformation compatibility.

To solve equation 5 both τ_{su} and q_{bu} must be evaluated, usually by laboratory or in situ testing of the bearing materials and/or correlations with parameters such as penetration resistance and physical soil/rock properties. Depending upon the design method adopted, the parameters of interest are shear strength, state of in situ stress and the deformability of the bearing materials.

<u>Ultimate Shearing Resistance of Decomposed Rock</u>: There is very little data available to directly relate the shear strength of the decomposed rock to τ_{su} and q_{bu}. However, the behavior of this material during laboratory testing replicates that of a heavily overconsolidated clay. As the decomposed rock represents a transition between soft rock and hard clay, it is of interest to compare existing relationships between the ratio of skin friction to the undrained shear strength (identified as the shear strength reduction factor (α)) and the undrained shear strength (c_u). Figure 7 represents the α vs c_u relationships developed by Horvath, et al (12) for soft rock and the range of similar data observed from numerous load tests in stiff to hard clays.

The data indicates that α for the very stiff to hard clays and the soft rocks are similar for c_u values within a range of about 2 to 5 tsf (0.2 to 0.5 MPa). As this is also the range of interest for the dense to very dense decomposed mica schist/gneiss, the median line of the Horvath α vs c_u relationship has been adopted for the dense to very dense decomposed rock. Thus, pending more definitive data, the ultimate skin friction can be estimated from Figure 7 in terms of the undrained shear strength as:

$$\tau_{su} = \alpha\, c_u \quad (6)$$

In application to dense to very dense decomposed rock, c_u should be reduced to accommodate degradation of τ_{su} should shaft movements exceed those required to develop the peak shearing resistance.

An alternate method used for stiff/hard clay is based on the effective stress approach proposed by Burland (13). In this application, τ_{su} is related to the in situ horizontal earth pressure and the effective friction angle of the soil (ϕ') as:

$$\tau_{su} = k_o\, \sigma'_{vo}\, \tan \phi' \quad (7)$$

where k_o and σ'_{vo} are the coefficient of earth pressure at rest and the effective overburden pressure, respectively. The difficulty with this approach is the characterization of k_o in the decomposed rock horizon and evaluating the effects of drilled pier of installation on k_o and ϕ'.

<u>Ultimate End-Bearing of Decomposed Rock</u>: As the dense to very dense decomposed mica schist/gneiss behaves as a very stiff to hard clay during shear, it appears reasonable to evaluate the ultimate end-bearing as for a cohesive soil. Consequently, q_{bu} is given as a function of the undrained shear strength (c_u) as:

$$q_{bu} = N_c\, c_u \quad (8)$$

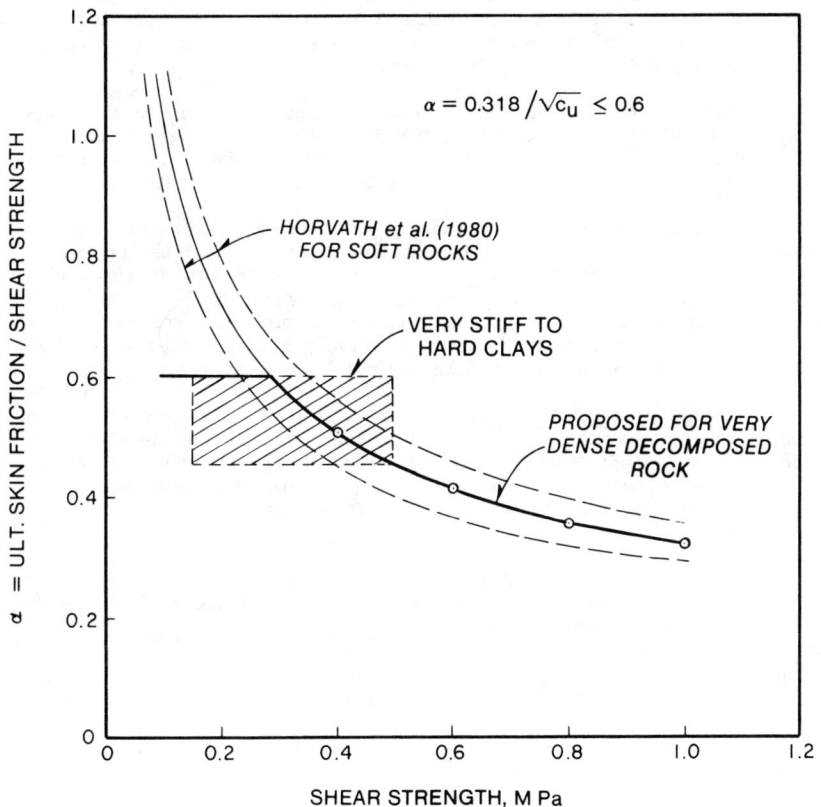

FIGURE 7 RELATIONSHIP BETWEEN THE UNDRAINED SHEAR STRENGTH OF DECOMPOSED ROCK FOR $N_{STP} >$ 100 BLOWS / 30 cms.

where N_c represents the bearing capacity factor. Expanding cavity solutions for N_c (14) yield values in the range of 6 to 12, depending upon the rigidity index (E_s/c_u). Alternatively, classic limiting equilibrium solutions from plastic theory indicate N_c = 9, which is the most commonly used factor for design. However, should the embedment of the base be less than 3 diameters, a linear interpretation of N_c should be made between 6 at the surface of the bearing materials and 9 at a penetration of 3 diameters.

For a dense to very dense decomposed rock with c_u in the range of 3 to 6 tsf (0.3 to 0.6 MPa), it follows that q_{bu} ranges from 27 to 54 tsf (2.5 to 5.0 MPa). This is consistent with regional experience where q_{all} values are typically 2-1/2 to 3 times the predicted q_{bu}.

Similar but slightly higher q_{bu} values are predicted using an effective stress approach. This approach requires evaluation of the bearing capacity factor (N_q) as a function of the ϕ' of the decomposed rock and the effective overburden pressure (σ'_{vo}). Compared to the undrained strength approach, there is somewhat greater uncertainty in characterizing ϕ' under relatively high stresses as well as σ'_{vo} where the piezometric profile may not be hydrostatic.

End-Bearing of Intact Rock: In many cases it is cost-effective to extend piers through decomposed rock horizon to develop end-bearing on intact rock. As the allowable end-bearing is almost always controlled by the compressibility of the rock mass rather than the ultimate bearing capacity, the deformation modulus of the rock mass (E_{rm}) is a key design parameter.

Since rock mass compressibility controls, design pressures for end-bearing have been established as a function of the rock mass modulus and the base settlement as:

$$q_{all} = \frac{E_{rm}}{\pi/4(1-\mu^2)} \frac{\delta_b}{d_b} \qquad (9)$$

If E_{rm} is related to the uniaxial compressive strength (q_u) and deformation modulus of rock cores, then:

$$E_{rm} = C_r q_u M \qquad (10)$$

where C_r is the E_{rm} correlation factor and $M = E_c/q_u$. Where q_u is interpreted from point load tests or where E_c has not been determined from uniaxial compression tests, M can be estimated from q_u vs M correlations made by Deere and Miller (15) for a variety of rock types.

To compare observed vs predicted behavior, q_{all}/q_u can be expressed by equation (11):

$$K = \frac{C_r M}{\pi/4(1-\mu^2)} \frac{\delta_b}{d_b} \qquad (11)$$

Comparisons of K predicted by equation (11) with Canadian Standards and the Australian Code of Practice are shown in the following table.

Table 1 - Ratio of q_{all} to q_u

Equation (11)[1]		Codes of Practice[2]	
$\delta_b/d = 0.5\%$	0.16-0.27	Canadian (16)	0.20-0.35
$\delta_b/d = 1.0\%$	0.31-0.52	Australian (17)	0.30-0.50

(1) M = 150 to 250; RQD < 60% (C_r = 0.15)
(2) Moderately close (0.3m) to wide (1.0m) joint spacing

Allowable end-bearing values have also been derived from experience at a number of locations in the U.S. (18). Those in the Atlantic Piedmont are often correlated with N_{SPT} and/or drill rig performance.

Load-Deformation Compatibility Analysis

A design analysis based on the premise of compatibility of load and associated deformation mitigates some of the uncertainty of the limit state design approach. The two design approaches described herein are elastic and load transfer analyses. Because of space limitations, only the elastic analyses method will be considered in some detail.

<u>Elastic Analyses</u>: This general method treats the soil/rock as an elastic continuum and has the facility to incorporate linearly varying and layered moduli, as well as to approximate nonlinear response. The work of Poulos and Davis (19), based on integral equation analysis, provides graphed solutions for a variety of pier dimensions and stiffnesses and is widely used.

The solution of Randolph and Wroth (20) is particularly noteworthy in that it can be applied with a hand calculator and provides significant insight to pier-soil interaction. This solution is based on a dimensionless "settlement ratio" (SR) defined by equation (12):

$$\frac{P_t}{r_o G_1 \delta_t} = \frac{\frac{4}{\eta(1-\mu)} + \frac{2\pi\rho\tanh(ul)}{\ln(r_m/r_o)ul}\frac{L}{r_o}}{1 + \frac{4}{\eta(1-\mu)}\frac{\tanh(ul)}{ul}\frac{1}{\pi E_p/G_1}\frac{L}{r_o}} \quad (12)$$

where;

$$ul = L/r_o \left[\frac{2}{\ln(r_m/r_o)E_p/G_1}\right]^{1/2} \quad (13)$$

$$r_m = L\left[0.25 + \left(2.5\,\rho\,(1-\mu)-0.25\right)\frac{G_1}{G_b}\right] \quad (14)$$

$$\eta = \lambda \frac{r_b}{r_o} \frac{G_1}{G_b} \qquad (15)$$

In equation (15), λ is reduced from 1 to 0.85 as q_{bu} is approached (22). The parameters in these equations are defined as:

E_p Modulus of shaft materials
G_b Shear Modulus of end-bearing materials
G_1 Shear Modulus at base of overburden
L Shaft length
P_t Applied load at top of shaft
r_b Radius of base
r_m Radius of influence of side shear stresses
r_o Radius of shaft
δ_t^o Settlement at top of shaft
μ Poisson's ratio
ρ Ratio of G at mid depth to G_1
η Correction for base diameter and bearing stiffness

The shear modulus may be varied with depth as shown by Figure 8. Two layer solutions can be approximated by a single G_1 representing the mean of the two moduli, weighted by layer thickness. Characterization of G_1 from pressuremeter tests is a preferred approach.

The shaft movement required to develop the peak shearing resistance (δ_z^*) is expressed from the Randolph and Wroth solution as:

$$\delta_z^* = \frac{r_o \tau_{su} \ln(r_m/r_o)}{G_1} \qquad (16)$$

The effect on δ_z^* of strain related degradation of G_1 radially from the shaft (increases δ_z^*) has also been considered by Kraft, et al (21). It follows that the elastic load limit (P_t^*) can be expressed as:

$$P_t^* = S_r \, r_o \, G_1 \, \delta_t^* \qquad (17)$$

P_t^* is the load at inception of nonlinear load-deformation response and represents the limit of validity for linear elastic analysis.

Provided $P_t = < P_t^*$, the load transferred to the base can also be expressed as:

$$\frac{P_b}{P_t} = \frac{4}{\eta(1-\mu)\cosh(ul)SR} \qquad (18)$$

Approximate methods are available for estimating the deformation of piers under loads beyond the elastic load limit (22) (19). The simplest approach, proposed for the integral equation solutions, assumes a trilinear simulation to the P_t vs δ_t curve as shown by Figure 9. The first linear segment is expressed as:

$$\delta_t = \frac{P_t}{r_o G_1 \, SR} \qquad (19)$$

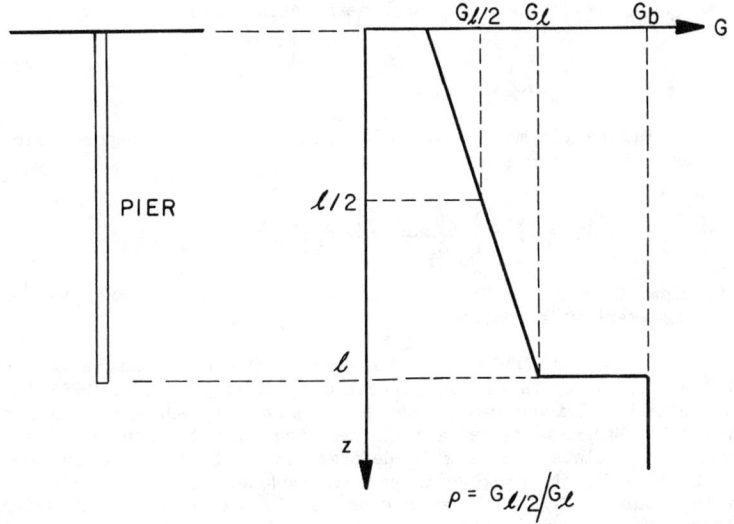

FIGURE 8 VARIATION OF SHEAR MODULUS WITH DEPTH

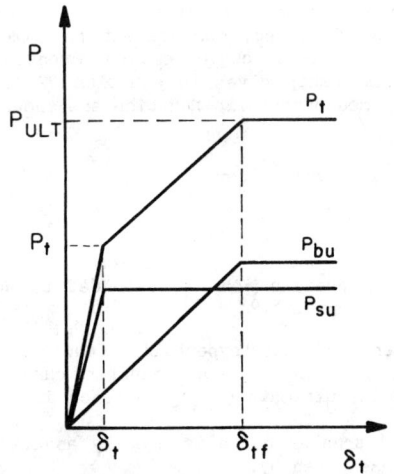

FIGURE 9 APPROXIMATION OF COMPLETE LOAD-DEFORMATION CURVE

If $\beta = P_b/P_t$, the inception of nonlinear response is:

$$P_t^* = \frac{P_{su}}{(1-\beta)} \tag{20}$$

where P_{su} is the ultimate shaft friction. The second segment simulates nonlinear response as:

$$P_t = \delta_t^* + \frac{(P_t - P_{su})}{r_o G_1 SR} + \left[(P_t - P_{su}) - \beta P_t^*\right]\frac{L}{AE} \tag{21}$$

and terminates at P_{ult}. Plunging of the pier is represented by continuous movement at a constant P_{ult}.

Load Transfer Analysis: To demonstrate the fundamentals of load transfer analysis, first suggested by Seed and Reese (23), the pier shaft is divided into a series of rigid segments as shown by Figure 10. The segments are joined together by linear elastic springs to simulate the elastic deformability of the segments under axial load. The shearing resistance vs movement of each shaft segment is simulated by nonlinear springs as is the base-load transfer. An iterative or closed form solution is implemented by assuming a small deformation at the base of the bottom shaft segment and solving for the mid-segment deformation and then the skin friction. This procedure is successively applied to each segment, finally determining the load and deformation applied at the top of the pier.

Shaft and base load transfer curves have been developed empirically from load test data and, more recently, from theoretical considerations (24) (21). These curves can be essentially replicated by the hyperbolic relationship given in equation 22 (25). The parameters of the transformed hyperbolic function are shown in Figure 11:

$$(\tau_s)_z = \frac{\delta_z}{\left(\frac{1}{K} + \frac{R_f \delta_z}{\tau_{su}}\right)} \tag{22}$$

where $\delta \le \delta_z^*$ and R_f is the τ_{hyp} correlation factor, usually taken as $0.85 \pm .05$. This expression can be extended to accommodate strain softening materials when $\delta_z < \delta_z^*$ (21).

The parameters of the hyperbolic function are the initial tangent modulus to the shaft load transfer curves (K), the shaft movement (δ_z), and the ultimate $(\tau_{su})_z$ for each discrete segment.

The base load tranfer curve is usually assumed to have an initial elastic response given by:

$$q_b = \frac{E_s \delta_b}{\frac{\pi}{4}(1-\mu^2)d_b} \tag{23}$$

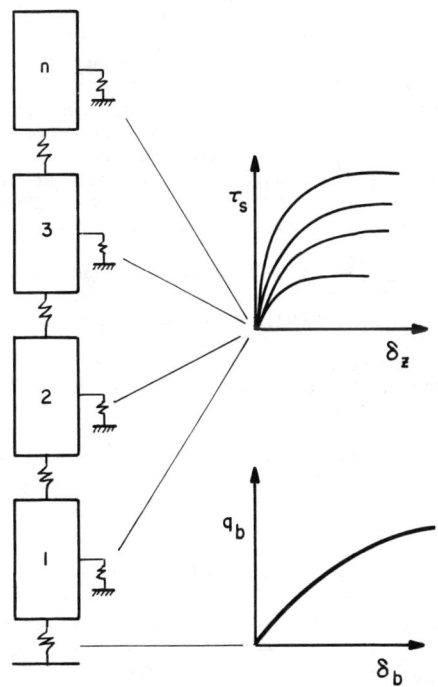

FIGURE 10 LUMPED MASS MODEL OF LOAD-TRANSFER METHOD OF ANALYSIS SHOWING SHAFT AND BASE LOAD TRANSFER CURVES.

Nonlinear response is usually assumed to initiate between 1/3 and 1/2 of q_{bu}. This response can be modeled similar to equation (22) by substituting δ_b and q_{bu} for δ_z and τ_{su}; and equating K to q_b/δ_b from equation (23).

This procedure has the advantage of simulating nonlinear P_t-δ_t behavior and is well suited to rapid parametric analysis, particularly in conjunction with a closed form solution incorporating equations (22) and (23). As the method does not consider the interaction effects of the load transfer from each shaft segment (except where the load-transfer curves are derived directly from load test), it is not exact. However comparisons with more exact solutions have indicated conformance, usually within 10 percent. Several axial load transfer computer programs are available in the public domain.

Rock Socket Analysis: Practice oriented solutions applicable to the design of rock sockets model the rock mass as a single homogeneous elastic half space (26), or as a two layer elastic media with different stiffnesses (27). For the latter case, the thickness of the upper layer corresponds to the socket length. Figure 11 depicts the finite element half space solutions of Pells and Turner (26) as a function of E_{rm}/E_ρ and the L/d of the pier.

To implement these solutions, the bond between the concrete shaft and the rock mass is expressed for smooth wall sockets as:

$$\tau_r = b\sqrt{q} \leq 1.7 \text{ MPa} \tag{24}$$

q is the weighted average of q_u (q_u^*) or the 28-day concrete compressive strength, whichever is less. The parameter b ranges from 0.20 to 0.25 as shown in Figure 7. As the rock mass quality is not considered in equation (24), it is prudent to reduce q_u^* to reflect rock mass defects as described for derivation of E_{rm} from E_c. It is significant that higher bond strengths can be mobilized with naturally rough or grooved socket walls (28).

A graphical design procedure (26), as an alternative to a trial-and-error solution, starts with selection of τ_r, E_{rm} and E_c. The maximum L/r_s of the socket is calculated from equation (25), assuming that there is no end-bearing:

$$\frac{L}{r_s} = \frac{P_{ts}}{2\pi r_s^2 \tau_r} \tag{25}$$

A line is drawn on Figure 12(a) between points at R = 100, L/r_s = 0, and R = 0, $(L/r_s)_{max}$. The intersection with the E_r/E_c curve gives P_b/P_t and the L/r_s of the socket. δ_{ts} is interpreted from equation (26) and Figure 12(b).

$$\delta_{ts} = \frac{I_b P_{ts}}{d_s E_s} \tag{26}$$

If either q_{all} or δ_{ts} exceeds design criteria, $(L/r_s)_{max}$ is increased and the design process is repeated as required. Where appropriate,

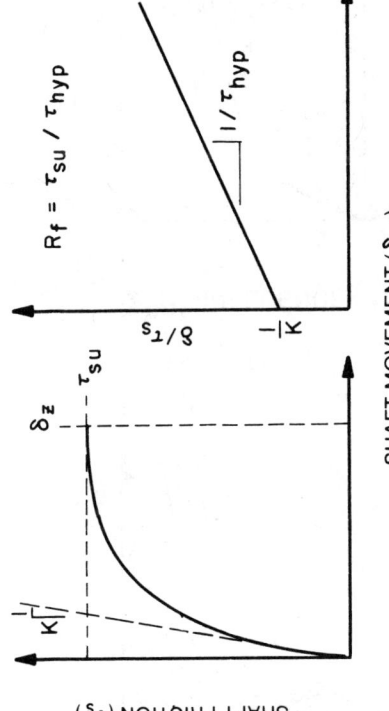

FIGURE 11 HYPERBOLIC AND TRANSFORMED HYPERBOLIC LOAD TRANSFER CURVE

82 FOUNDATIONS AND EXCAVATIONS

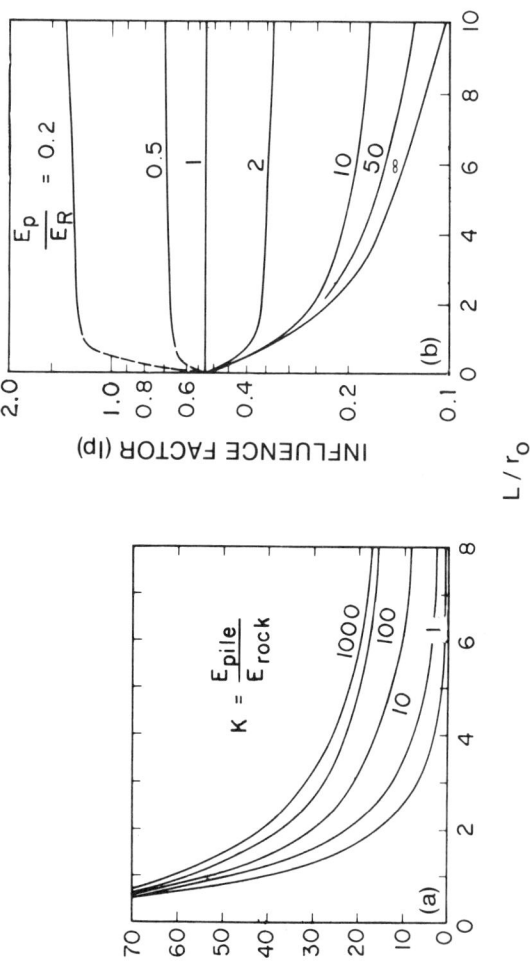

FIGURE 12 (a) LOAD TRANSFERRED TO BASE AND (b) SETTLEMENT INFLUENCE FACTORS [ ]

this solution can be coupled with the preceding elastic analysis (22) to estimate the load transferred to the socket.

DESIGN ANALYSES

To demonstrate design applications, an elastic analysis is conducted for a load pier previously designed by the load-transfer method. This straight-shaft pier was drilled 85 feet into a predominantly decomposed rock zone and comparatively little additional support was anticipated from the overburden. The parameters for elastic analysis are summarized below.

Pier Parameters		Geotechnical Parameters	
Design Load	2300 tons (2.04 MN)	G_1	1800 tsf (172 MPa)
Diameter	5.5 ft. (1.7 m)	G_b/G_1	2.30
Length	85 ft. (25.9 m)	ρ	1.0
Shaft Modulus	2.6×10^4 tsf $(2.5 \times 10^4$ MPa)	c_u(shaft)*	4.45 (0.43 MPa)
		c_u(base)	7.4 tsf (0.71 MPa)

*minimum ultimate shear strength = 3.67 tsf (0.35 MPa)

Ultimate Capacity: Assuming the minimum ultimate strength to accommodate strain softening effects, α = 0.6 from Figure 7 and τ_{su} = 2.2 tsf (0.21 MPa) from equation (6). For a bearing capacity factor of 9, q_{bu} = 66.6 tsf (6.38 MPa) from equation (8). The ultimate capacity of the pier is then 4770 tons (457 MN).

Elastic Response: For the geotechnical data given, the settlement ratio from equation (9) is calculated as 29.69. The elastic response is:

$$\delta_t = 8.15 \times 10^{-5} P_t \text{ (inches)} \qquad (27)$$

Based on β = 0.123 (equation 14) and the minimum average post-peak skin friction of 2.2 tsf (0.21 MPa), the load at the elastic limit (P_t^*) is about 3,600 tons (32 MN) from equation (17) and the corresponding δ_t^* is 0.3 in. (0.76 cm).

Inelastic Response: For loads > P_t^*, non-linear load-deflection is approximated by calculation of the deformation at the ultimate load by substituting P_{ult} for P_t in equation (18). At the ultimate load of 4770 tons (457 MN), yield (plunging) occurs at an estimated deformation of 1.5 inches (4.6cm). The trilinear P_t vs δ_t response shown in Figure 13 is superimposed on the pier load-deformation curve predicted as part of a previous design study. The elastic response provides a good correlation up to about one-half of the ultimate load where the nonlinearity of the P_t-δ_t curve begins to rapidly increase. The unusually small deformation at ultimate load is consistent with the assumption that there may be relatively small strains required to cause base failure of the rock-like bearing materials.

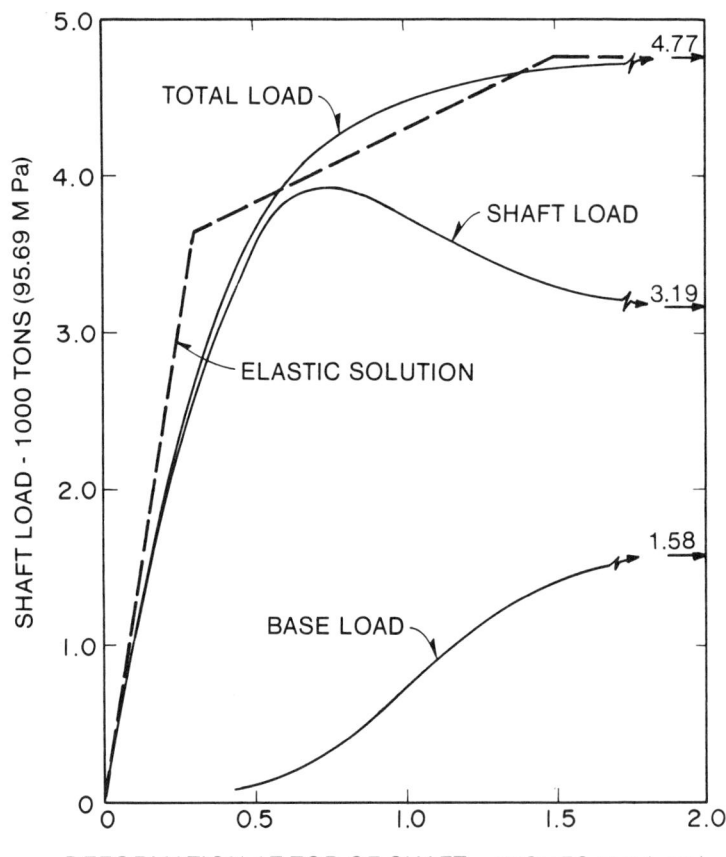

FIGURE 13 COMPARSION OF ELASTIC SOLUTIONS WITH LOAD-TRANSFER ANALYSIS OF 5.5 FT. (1.7 m) SHAFT DRILLED 85 FT. (25.9 m) INTO DECOMPOSED MICA SCHIST

References

1. Sowers, G.F., (1954), "Soil Problems in Southern Piedmont Region", Proc. ASCE, J. Soil Mech. Found. Engrg. Div., Separate No. 416.
2. Hall, W.J., Newmark, N.M. and Hendron, A.J., (1974), "Classification, Engineering Properties and Field Exploration of Soils, Intact Rock and In Situ Rock Masses", Wash-1301, Directorate of Regulartory Standards, U.S. Atomic Comm., Washington, D.C.
3. Tarkoy, P. (1975), "A Study of Rock Properties and Tunnel Boring Machine Advance Rates in Two Mica Schists Formations", Applications of Rock Mech., Proc. 15th Symp. on Rock Mech., ASCE.
4. Hunt, R.E., (1984), "Geotechnical Engineering and Investigation Manual", McGraw-Hill Book Co., New York.
5. Deere, D.U., Hendron, A.J., Patton, F.D., and Cording, E.J., (1967), "Failure and Breakage of Rock", Proc. 8th Symp. on Rock Mechs., Amer. Inst. of Mining and Metallurgy Engrg., Minneapolis, MN.
6. Bieniawski, Z.T., (1978), "Determining Rock Mass Deformability", Intl. J. Mech. Min. Sci., Vol. 15, No. 5.
7. Coon, R.F. and Merritt, A.T., (1969), "Predicting In Situ Modulus of Deformation Using Rock Quality", ASTM, Special Tech. Pub. 477.
8. Bieniawski, Z.T., (1974), Geomechanics Classification of Rock Masses and Its Application to Tunneling", Proc. 3rd Intl. Cong. Rock Mech., Vol. II A, Denver, CO.
9. Chappell, B.A. and Maurice, R., (1980), "Classification of Rock Mass Related to Foundations", Proc. Intl. Conf. on Structural Founds. on Rock, Univ. of Sydney.
10. Osterberg, J.O., (1984), "A New Simplified Method for Load Testing Drilled Shafts", Found. Drilling, Intl. Assoc. of Found. Drilling Contractors, Dallas, TX.
11. Kulhawy, F.H. and Goodman, R.E., (1980), "Design of Foundations on Discontinuous Rock", Proc. Intl. Conf. on Structural Found. of Rock, Univ. of Sydney.
12. Horvath, R.G., Kenney, T.C. and Kozicki, P., (1983), "Methods of Improving the Performance of Drilled Piers in Weak Rock", Canadian Geot. J., Vol. 20, No. 4.
13. Burland, J.B., (1973), "Shaft Friction of Piles in Clay - A Simple Fundamental Approach", Ground Engrg, Vol. 6, No. 3.
14. Vesic, A.S., (1977), "Design of Pile Foundations", NCHRP Synthesis of Hwy. Prac., No. 42, Transp. Research Board, Washington, D.C.
15. Deere, D.U. and Miller, R.P., (1966), "Classification and Index Properties for Intact Rock", Tech. Report AFWL-TR-65-116, AF Special Weapons Center, Kirtland Air Force Base, New Mexico.
16. Canadian Geotechnical Society, (1978), "Canadian Foundation Engineering Manual", National Research Council, Canada.
17. Standards Assoc. of Australia, (1975), "The Design of Installation of Piling", Draft Document DR75020 and Supplements.
18. Woodward, R.J., Gardner, W.S. and Greer, D.M., (1972), "Drilled Pier Foundations", McGraw-Hill Book Co., New York.
19. Poulos, H.G. and Davis, E.H. (1980), "Pile Foundation Analysis and Design", John Wiley & Sons, New York.

20. Randolph, M.F. and Wroth, C.P., (1978), "Analysis of Deformation of Vertically Loaded Piles", J. Geot. Engrg. Div., ASCE, Vol. 104, GT-12.
21. Kraft, L.N., Ray, R.P. and Kagawa, T., (1981), "Theorectical t-z Curves", J. Geot. Engrg. Div., ASCE, Vol. 107, GT-11.
22. Randolph, M.F. and Wroth, C.P., (1978), "A Simple Approach to Pile Design and the Evaluation of Pile Tests", ASTM, Special Tech. Pub. 670.
23. Seed, H.B. and Reese, L.C., (1957), "The Action of Soft Clay Along Friction Piles", Trans. ASCE, Vol. 122.
24. Vijayvergiya, V.N., (1977), "Load-Movement Characteristics of Piles", Proc., Ports 77, ASCE, Vol. 2.
25. Donald, I.B. and Chiu, H.K., (1980), "Theoretical Analysis of Rock-Socketed Piles", Proc. Intl. Conf. on Structural Found. on Rock, Univ. of Syndey.
26. Pells, P.J.N. and Turner, R.M., (1979), "Elastic Solutions for the Design and Analysis of Rock-Socketed Piles", Canadian Geot. J., Vol. 16, No. 3.

DRILLED PIERS IN THE PIEDMONT - MINIMIZING CONTRACTOR-ENGINEER-OWNER CONFLICTS

Stanley A. Schwartz*
Member, ASCE

ABSTRACT: Drilled piers are used extensively in the Piedmont Physiographic Region, deriving support primarily from end-bearing on hard rock. Cost extras and contractor-engineer-owner conflicts are common because of poor design, specifications and contracts; variable subsurface conditions; uncooperative and/or inexperienced contractor personnel; and unpredictable and/or unreasonable construction monitoring demands. This paper is intended to describe these problems and present recommendations for mitigating them.

INTRODUCTION

In the Piedmont, heavily loaded structures are generally supported by drilled piers or one of several pile types. The choice of drilled piers or piles is typically one of comparative costs. The cost of a drilled pier foundation is usually the more difficult to predict because of the greater impact that variable conditions and construction monitoring have on cost. Poor specifications, related to the drilling equipment, procedures and bearing level, can also result in large cost extras. Consequently, if the certainty of front-end contract cost is valued, rather than minimum base bid contract cost with risk of extras, drilled piers are not looked upon as favorably as piles. Association of Drilled Shaft Contractors (ADSC) is keenly aware of this perception and has actively participated with both member contractors and the A/E community to develop ways to mitigate the problem. Geotechnical engineers can play a significant role also; it is in their best interest to participate heavily in the development of specifications and contract language so that drilled piers can be utilized, when appropriate, and construction can be accomplished with minimum disputes, delays and cost extras.

In the Piedmont, drilled piers are used primarily for foundation support of heavy structures, typically more than six stories in height. In the southern Piedmont, straight shafts are typically extended down to hard rock to achieve primary support in end-bearing, and bells are often constructed manually. The piers are constructed in holes temporarily cased with steel liners. The bottom is cleaned of loose material and water, and then evaluated by a geotechnical engineer to confirm the design bearing pressure or recommend alternative construction measures.

In typical Piedmont practice, a large rig, drilling with a rock auger, will refuse on rock that is hard to very hard. If intact, the rock would exhibit unconfined compressive strengths far in excess of that of concrete. However, design bearing pressures are controlled by the presence of joints, fractures and seams filled with soil or partially weathered rock. During the

*Senior Vice President, Soil & Material Engineers, Inc., 3300 Marjan Drive, Atlanta, GA 30340

past 30+ years of drilled pier design and construction in the Piedmont, maximum design bearing pressures have increased from 50 to 60 ksf (2.4 to 2.9 MPa) to 100 to 150 ksf (4.8 to 7.2 MPa). This is due mainly to the use of much larger, more powerful drilling equipment, and better cutting tools which penetrate softer fractured rock to much harder continuous rock.

The success of a drilled pier foundation project is dependent on several factors, many of which can directly involve the project geotechnical engineer. Initially, the geotechnical engineer must obtain sufficient subsurface data to establish that drilled piers are suitable and economical for the project. He must work closely with the owner and A/E design team to assist in formulating a safe, practical and cost-effective design, and a clear specification which properly accounts for established local practice in the construction and monitoring of drilled piers. He should encourage and participate in pre-bid and pre-award conferences with drilled pier contractors. After selection of a qualified drilled pier contractor, a pre-construction meeting should be held, attended by the A/E, general contractor, drilled pier contractor and the project geotechnical engineer and designated inspector. Procedures involved in drilled pier construction and monitoring should be reviewed and all parties should understand their respective roles for successful construction of all drilled piers.

Unfortunately, on many construction projects, owners and drilled pier contractors become involved in litigation, with the lawyers coming out as the only winners. Great strides have been made in the past two to three years in attacking this problem through the cooperative efforts of the Deep Foundations Institute (DFI), ADSC and representatives of American Society of Civil Engineers (ASCE), Association of Soil and Foundation Engineers (ASFE) and other interested professional industry groups. The Deep Foundation Construction Industry Roundtable (DFCIR) has been the primary forum of these efforts. A "100 Day Document" was recently produced. This agreement, to be applicable to all involved parties on a project, details sequential mandatory discussion, mediation and arbitration for resolution of all disputes. It is hoped that this agreement, when used, will avoid many contract dispute claims.

DESIGN OF DRILLED PIERS

Selection of Geotechnical Consultant - An important first step toward the success of foundation design and construction on a project is the selection of a qualified, experienced geotechnical engineering consultant to undertake a detailed subsurface exploration and evaluation. PAESAR (Professional Architect-Engineer Selection and Retention), formulated by ASFE, should be used in selection of the geotechnical engineering firm. This method involves selection based on qualifications and commitment to the project.

Too often, geotechnical firms are asked to compete on a price basis; this can, and often does, result in a less thorough investigation than is appropriate. Subsurface data are less extensive and the geotechnical services are of lower quality. In a highly competitive environment, where the owner or A/E implies or states restrictions on exploration costs, positive interaction and involvement of the geotechnical engineer with the design team is limited; this limit of communication often results in a poor design and specification. A greater likelihood also exists that unnecessary extras or conflicts between the contractor and geotechnical engineer-design A/E-owner will occur. The bottom line is a more expensive foundation.

The drilled pier contractor can and should assist the geotechnical engineer, when appropriate, by requesting of the general contractor and design team that sufficient subsurface data be available for formulation of a reasonable bid.

Geology and Subsurface Conditions of the Piedmont - To plan an adequate subsurface exploration, the geotechnical engineer must be knowledgeable of the general and specific geology of the project site. The geotechnical engineer's experience with prior explorations and deep foundation construction within the general area and within similar geologic settings is invaluable. Lack of experience and/or lack of a thorough exploration and proper interpretation can lead to difficulties and conflicts during construction when unexpected conditions are uncovered. This is particularly true in the Piedmont, which is characterized by highly variable subsurface conditions.

The Piedmont Physiographic Region extends for over a thousand miles, from central Alabama trending northeasterly through Georgia, eastern Tennessee, South and North Carolina, and northerly through Virginia and Maryland. A map of the Piedmont region is presented as Figure 1 below:

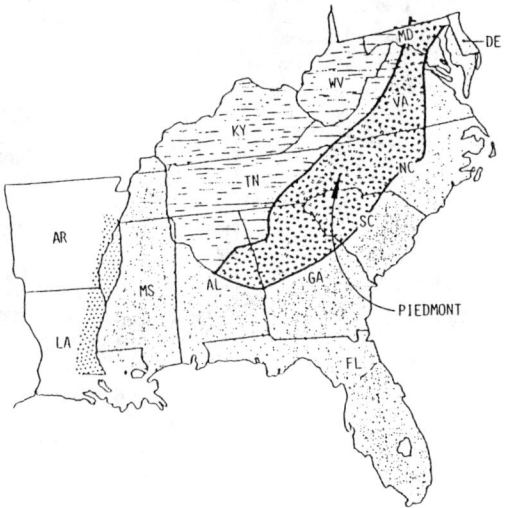

Figure 1. Map of Piedmont Physiographic Region

The Piedmont contains the oldest rocks in the southeastern United States. These rocks are more than 600 million years old. The primary rock types within the Piedmont are sedimentary and igneous rocks which have been subjected to repeated cycles of metamorphism, folding, faulting and intrusion. In response to the various stresses, the rock mass contains numerous cracks, joints and planes of weakness. Weathering is highly variable. Gneiss and schist are the principal rock types, with sandstones and siltstones scattered throughout the region.

Throughout the Piedmont, bedrock is overlain by residual weathered products of the parent rock. Over a period of hundreds of millions of years, the rock has been chemically and physically altered by the elements, producing, in many areas, an upper zone of clayey soils underlain by sandy silts and silty sands. The residual soil zone is typically 20 to 50 feet thick (6 to 15 m), but sometimes more than 70 feet (21 m) thick. Partially weathered rock (PWR) is usually present as a transitional material between the overburden soils and the underlying unaltered bedrock. Partially weathered rock is defined as material exhibiting standard penetration resistance values in excess of 100 blows per foot (per 0.3 m). The PWR can be as thin as a few feet and as thick as 50 feet (15 m) or more.

Bedrock is generally defined as material exhibiting a standard penetration resistance of 50 blows with no penetration and causing refusal to a solid-stem or hollow-stem auger or drag bit. Unfortunately, there is no standard definition of refusal, and reported refusal levels may vary depending on which firm, drill rig, or driller is used to obtain the subsurface data.

A typical Piedmont subsurface profile is presented as Figure 2 below:

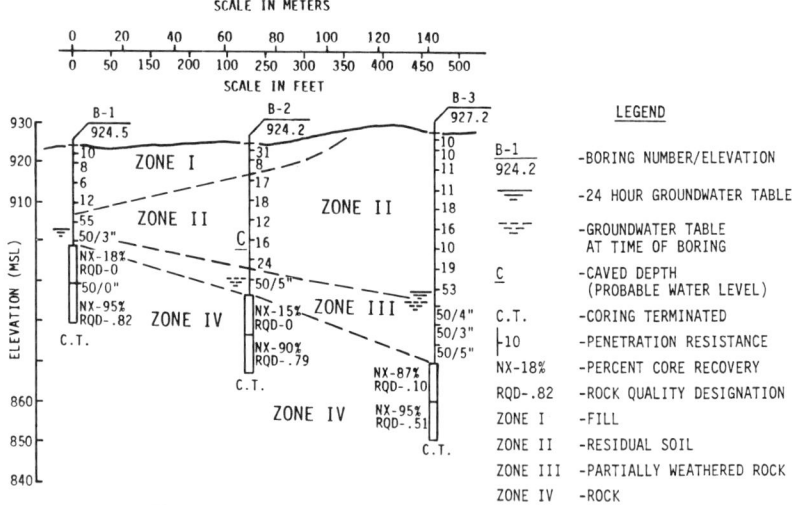

Figure 2. Typical Piedmont Subsurface Profile

The rock below the refusal level often contains fractures, seams and joints for several feet until more unaltered rock is reached. Of significance, some Piedmont sites contain hard rock underlain by zones of soil or partially weathered rock before more massive rock is again reached. The presence of these rock anomalies and the variation in reported refusal levels are significant factors in the design and costs associated with drilled pier construction. The actual strength of the rock within the Piedmont seldom, if ever, dictates the allowable (design) bearing pressure. Unconfined compressive strengths may vary from 500 to 1000 psi (3.4 to 6.9 MPa) for some schist and softer gneiss to in excess of 10,000 psi (69 MPa) for hard gneiss, granite or quartzite. Design pressures utilized are often only 10 to 20 percent of the unconfined compressive strength of the intact rock.

Subsurface Exploration - The number of borings and amount of rock coring required to properly evaluate subsurface conditions for a given project is dependent on several factors. Because the Piedmont has highly variable subsurface conditions, an accurate prediction of the conditions to be encountered at specific column locations is not possible. Interpolation of data from widely spaced borings can result in very misleading conclusions, and constructed foundation depths can differ widely from those anticipated.

For heavy structures, it is desirable that borings be spaced no more than about 100 feet (30 m) apart. If a lump sum foundation contract is desired and/or the owner wants extras to be at a minimum, it would be desirable to perform borings at every other column location, or on approximate 60- to 75-foot (18 to 23 m) centers. On projects where drilled piers are considered as a probable foundation type, it is desirable that the four building corners and one central boring be cored. In addition, borings should be drilled at the shear wall locations to provide data for an evaluation of uplift resistance. Rock coring should ideally be to a minimum of 10 feet (3 m). The final core run should typically have a recovery of at least 70 percent and an RQD (rock quality designation) of at least 50.

During rock coring, close attention should be paid to the condition of the drill bit, the down pressure used, and the time of penetration required to advance the core barrel through the rock. The rock should be very carefully boxed, then closely examined and stratified. The percent recovery, RQD, and defects (closed or open) and other details should be accurately described. At least 12 hours after drilling completion, a groundwater measurement should be made and reported.

The geotechnical engineer should present all the subsurface data clearly on soil and rock boring records, so that foundation contractors can evaluate conditions likely to be encountered during construction. Cross-sections and contour plans of estimated rock and groundwater elevations are very helpful visual aids. However, because of the high variability of the Piedmont rock, the geotechnical engineer should include in his report warning statements regarding the possible variation of conditions interpolated between data points.

Evaluation of Subsurface Data and Development of Design and Specs - Once a clear subsurface model has been developed for the site, the evaluation of drilled piers versus piles can be performed. The geotechnical engineer should share his evaluation in an open forum with the owner and A/E team. When the general contractor is preselected, he too should be included. This fosters an environment for open discussion and understanding of the pros and cons of drilled piers versus piles.

Several drilled pier designs are possible within the Piedmont, depending primarily on column loads and subsurface conditions. Heavy structures would typically be supported by drilled piers extended down to hard rock below the water table. Construction of these piers requires casing, down-hole pumping, cleaning, evaluation of the bottom by the on-site engineer and dry concreting operations. Design bearing pressures for piers in the Piedmont have ranged from 20 ksf (1 MPa) within partially weathered rock to 150 ksf (7.2 MPa) on very hard continuous rock. The assigned bearing pressure is dictated by settlement concerns rather than by the actual bearing capacity of the rock.

The geotechnical engineer should work in conjunction with the A/E designers and general contractor (when preselected) in developing preliminary cost estimates of drilled pier and pile alternatives. To design and specify the drilled pier, several questions must be addressed. One of the primary decisions is whether the drilled piers should be constructed with straight shafts, utilizing only a small percentage of the available concrete strength but minimizing costly rock excavation in constructing bells. The alternative is to construct the smallest shaft that the codes allow, and accept that belling in rock may be required to achieve sufficient end-bearing area. Trial designs can be formulated by the structural engineer for unreinforced concrete (nominal 0.5 percent reinforcing usually used), which permits the stress within the concrete up to limits of about 25 to 32 percent of design compressive strength (f'c), and for reinforced concrete, which permits an allowable stress range of about 35 to 50 percent of f'c (steel varying from about 1 to 3 percent).

Reputable local drilled pier contractors should be consulted on the approximate prices of shaft and bell construction. In Atlanta, the shaft is priced at about $200 per yard (per 0.7 m^3), with a probable range on most jobs between about $175 and $250 (not including steel). Removal of rock, to construct a bell or to extend the shaft below the rock auger refusal level, costs about $1,000 to $1,500 per yard (per 0.7 m^3). Trial calculations will readily demonstrate that for pier lengths of more than about 20 to 25 feet (6 to 8 m), pier design is optimized by using the smallest diameter shaft possible, to save considerable quantities of shaft concrete, and accepting the high cost for rock excavation to construct the bell.

Final design for a project should involve only a few drilled shaft sizes, as this tends to reduce the overall cost. Thus, it may be desirable to slightly increase or decrease the design bearing capacity for specific piers to minimize the number of different diameter shafts. If the column loads are relatively light, the design bearing capacity may be computed as the load divided by the area of a 30-inch (0.8 m) diameter shaft, as this is the smallest shaft diameter that can be practically hand-worked and evaluated.

CONTRACT AND SPECIFICATION DEFICIENCIES

Once design is finalized, the next challenge for the geotechnical engineer is to assist the A/E team in preparing a project-specific specification-contract bid package for selection of an experienced, reputable drilled pier contractor. Several good documents are available that provide sample specifications for use as guides. These include the ACI Standard 336.1-79, revised 1985; ACI 336.3R-72, reaffirmed 1980; the Drilled Shaft Manual (Lymon C. Reese); and Drilled Pier Foundations (Woodward, Gardner and Greer). However, even a very good guide specification typically must be modified to meet the needs of a specific project.

On projects with poor contracts, plans and specifications, disputes are quite likely. Unaltered AIA documents should be incorporated in all contracts to minimize disputes. A mandatory mediation-arbitration clause should also be included, so that disputes can be resolved rapidly and judiciously. Specifications must be flexible and not transfer unfair risks to the drilled pier contractor.

The relationship between the geotechnical engineer and the A/E is important in developing appropriate specifications. If this relationship is weak, or if monies are not available to permit the geotechnical engineer to at least review the proposed specifications, a poor specification and contract are more likely to result. There is then the consequent likelihood of contractor conflicts and large cost extras. Too often, the specification is "pulled off the shelf" and is largely inappropriate for the intent of the design. In preparation for this paper, more than a dozen specifications were reviewed; numerous requirements which are unnecessarily severe, impractical, unfair and, in some cases, technically incorrect, were found. The most significant of these specification requirements will be reviewed subsequently.

Among the cost extras on drilled pier projects, the two most significant are the variation in pay length and rock quantities from the base bid quantities. Influencing these variations are three primary factors: 1) the geotechnical engineer or structural engineer must estimate the level that suitable bearing can be achieved. Bearing levels are estimated initially at the boring locations and then interpolated to column locations where piers will be constructed. Because of the highly variable conditions in the Piedmont, it is difficult to accurately estimate the bearing levels; 2) the geotechnical engineer must estimate the level to which the drilled pier contractor can advance the shaft by mechanical means, and, consequently, the volume of additional shaft and/or bell construction which will be classified as rock excavation. In addition to the inherent difficulties in making this estimation, the refusal level achieved by the contractor is also dependent on the condition and type of equipment, the tools used, the operator, and the specification requirement and/or willingness of the contractor-operator to slowly grind on soft or fractured/seamy hard rock prior to terminating shaft advancement with the rig. It is noted, too, that the contractor-operator can falsely make it appear that rig "refusal" has been reached, above the level of true refusal. The on-site geotechnical engineer often cannot detect the deception; 3) the third factor is the variable judgment in the field by the on-site engineer. The decision of whether the rock at the base of a drilled shaft is suitable for the design pressures cannot be made quantitatively, with exactness. The drilled pier contractor, and ultimately the owner, is subjected to unknown and variable degrees of conservatism by the on-site engineer in deciding to advance the shaft further with rock removal methods or to hand-construct a bell at or below the rig refusal level. In addition, the on-site engineer is often not the geotechnical firm's most experienced engineer relative to drilled pier evaluations.

It is desirable, but difficult, to write a specification which assures that the pier will be drilled to "exactly" the required bearing level, rather than several feet above or below it. If the geotechnical engineer could be reasonably certain of the required bearing level at each foundation location, specifications could be written on a performance basis; these would require the contractor to achieve this specific level, using whatever means necessary, and would grant no extras for the equipment, tools or techniques used. However, without drilling one boring at each foundation location, it is not possible to accurately predetermine bearing levels; and even by drilling at each location, there would be no guarantees. Variations from anticipated bearing levels are inevitable. This limitation aside, several projects in Atlanta have been bid and successfully completed on a performance "lump sum" basis. The owner likely paid a premium on these projects for up-front certainty of foundation cost. This contract type puts a large burden on the geotechnical engineer/on-site engineer, since the drilled pier contractor will be ever watchful and concerned that the on-site engineer is overly conservative in his construction demands.

The issue of whether to specify minimum equipment, tools and experience to better assure that shafts do not refuse prematurely above bearing rock, is a difficult one. Drilling equipment and tools vary widely; therefore, specifying minimums may exclude a reputable contractor who can satisfactorily complete the work with out-of-spec equipment, by using better tools and methods, or a more experienced operator. A suggestion of ADSC is that the probable acceptable level and type of rock for bearing be clearly stated, and the responding drilled pier contractors be required to include a description of the equipment, tools and methods which will be used to achieve these levels. The burden is then on the geotechnical engineer in advising the A/E, owner and general contractor on the relative merits of the competing contractors' bids. An open interview with the drilled pier contractors, in the presence of the A/E, owner and general contractor, could be beneficial in expressing and resolving questions concerning their submittals and bids. At present, this writer holds to the opinion that the specification should require certain minimums related to the experience of the contractor and personnel who work on the project, and to the equipment and designated tools to be used. The specification wording should be such that all reputable contractors who have satisfactory (though varying) equipment can bid the project.

Some other of the more controversial requirements often included in drilled pier specs follow:

1. **Full or Part-Time Construction Monitoring** - Many specifications require that the drilled pier contractor "notify the architect and testing agency at least 24 hours prior" to the time excavations will be ready for evaluation. Drilled pier contractors feel that full-time construction monitoring is mandatory on all jobs where more than one hole is being opened and worked at a time. The writer strongly agrees.

2. **Classification of Rock** - ADSC feels that rock should be defined as any material which cannot be drilled with a conventional "earth auger". This would be appropriate for piers designed for only moderate end-bearing capacity of 40 to 60 ksf (2 to 4 MPa). In the writer's opinion, piers designed to bear on 100+ ksf (5 MPa) rock should be advanced to "rock auger" refusal, where refusal is defined as a penetration rate of less than 6 inches (15 cm) in 15 minutes with the minimum specified rig. Below the rock auger refusal level, rock extras would apply. This eliminates one more element of potential controversy and dispute by not having to differentiate "earth auger" and "rock auger" levels for determination of pay quantities.

3. **Refusal Above Bearing Level** - The following sketch (Figure 3) illustrates two typical situations where refusal is encountered above the eventual bearing levels. In these cases, all specified requirements for advancing the shaft with mechanical means are satisfied, but refusal is encountered either on a "float boulder" in the overburden or on sloping

bedrock near final bearing. If all subsequent excavation below initial refusal is classified as rock (as recommended by ADSC, and often specified) very large extras, not totally deserved by the contractor, will result. As illustrated by Figure 3, a considerable part of the subsequent excavation volume is in soil, that can be removed with much less effort than can rock. What is fair compensation to the contractor and not unfair to the owner? In the case of the boulder, the contractor is delayed from the normal progress of advancing the shaft. His equipment must either stand-by idle at this location, until the obstruction is removed, or remobilize to this location after the hand-removal is completed. In the writer's opinion, a fair compensation is to credit the contractor with an extra for removal of the obstruction equal to two times the actual volume of the rock removed (two times A) or the volume corresponding to the height (Depth 1) times the cross-sectional area (A plus B), whichever is less. Where there is sloping rock at the base, rock should be removed to form a slope not steeper than typically 15%, or steps constructed with heights not greater than typically 25% of the pier diameter. The full volume corresponding to the height (Depth 2) removed, times the cross-sectional area, should be credited to the contractor.

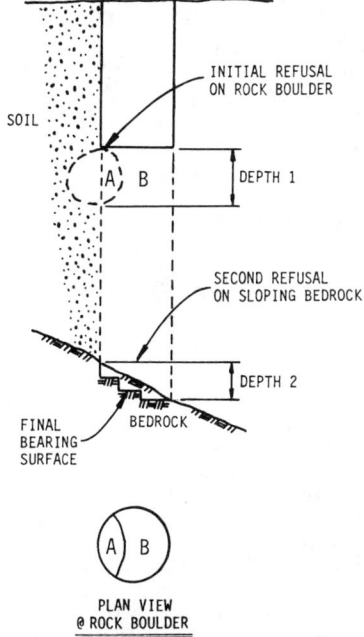

Figure 3. Typical Premature Refusal Conditions

4. **Calculation of Shaft and Rock Extras** - Specifications and bid documents are rarely clear as to whether the total shaft length ("pay length") is inclusive or exclusive of the shaft length paid for as a rock extra. These measurements are illustrated in Figure 4 below. Either method is acceptable as long as it is clear to the drilled pier contractor and other interested parties.

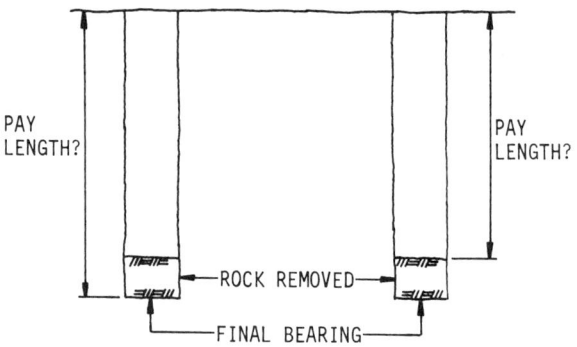

Figure 4. Pay Length Variation

5. **Plumbness** - A plumbness tolerance of 1% is sometimes specified. This is considered unnecessarily restrictive in typical cases. Sloping rock and boulders often cause the drilling tools to "kick-out" of plumb, and extreme measures are often necessary to satisfy a 1% plumbness criterion. A 1½ or 2% tolerance is more standard in the industry and should be accommodated by design whenever possible.

6. **Unauthorized Excavation** - Specifications often define unauthorized excavation as material removed beyond bearing elevations indicated on the plans or column schedules. On most projects, where it is intended that the contractor extend the pier to rock auger refusal, this specification is redundant and incorrect. However, if the design intent is that piers bear instead within partially weathered rock, defining unauthorized excavation could be a very necessary part of the specifications. Some projects actually have this design intent, yet the specification requires that the contractor use a very large drill rig and advance the shafts to rock auger refusal. This could result in shaft lengths considerably in excess of the design estimates corresponding to termination in partially weathered rock. This is an inconsistency which, if not realized prior to finalization of the contract with the drilled pier contractor, can lead to a huge extra, considerable embarassment, and probable litigation between the owner and the A/E-geotechnical engineer team.

7. **Pier Shafts Not Left Open For More Than 24 Hours** - Several specifications require that the drilled shaft construction be completed and concreted within 24 hours and/or that the construction not be interrupted by weekends, holidays or for any other reason without the specific approval of the geotechnical engineer. This specification may have merit in special circumstances when leaving a hole open may incur risks to adjacent construction or when shaft resistance in PWR is considered. It is only in such situations, however, that specifications should so restrict the contractor in the execution of his work. If a hole is cased properly and proper protection measures are installed at the surface, there is no technical reason why a shaft could not be left unworked for an extended period, if desired by the contractor.

8. **Shoring Bells** - Some specifications require that bells be shored, and that the shoring be left in place unless otherwise directed by the engineer. Seldom, if ever, is shoring necessary in the Piedmont when bells are constructed within the rock mass. If the bells are in soil or partially weathered rock below the groundwater table, collapse is possible. Solutions preferable to shoring are to extend the straight shaft to more competent material or to bell at a lower level in more competent rock. In no case should the geotechnical engineer be responsible for **directing** the contractor related to safety aspects involving shoring.

9. **Belling Above or Below Rock Auger Refusal** - Belling above rather than below the level of rock auger refusal can be accomplished faster and for a lesser cost extra. This is illustrated in Figure 5. To enable belling above rock auger refusal, the temporary casing must be set a few feet above the rock auger refusal "bottom". The specification should encourage, but not require, this. Only the contractor's personnel should judge whether this can be done while still providing a proper seal against excessive water inflow.

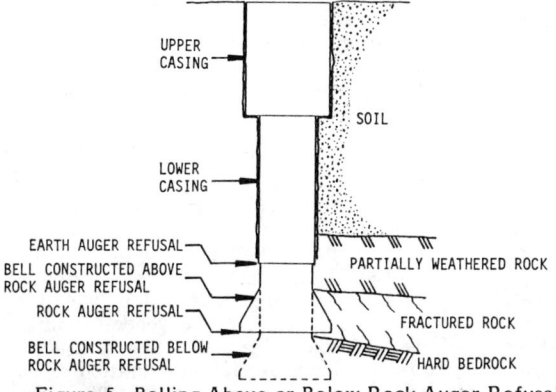

Figure 5. Belling Above or Below Rock Auger Refusal

10. **Rock Removal by Blasting** - Specifications are often not clear in permitting or excluding the use of blasting methods for rock removal. If blasting is permitted, the contractor should be required to submit blasting plans to the design engineer; and the engineer should reserve the right to restrict blasting to certain maximum charges and delays so as not to damage intact rock or recently poured concrete in adjacent piers.

11. **Concrete Specification** - Inadequacies in drilled pier construction historically have involved poor concreting operations more than any other single factor. A "good" specification and thorough construction monitoring related to concreting are therefore very important. A qualified professional materials engineer, experienced in "good" drilled pier concrete specifications and procedures, should be consulted during formulation of the specifications.

Concrete slump, an item often in controversy, should typically be in the 5- to 7-inch (13 to 18 cm) range rather than 4 inches (10 cm) as is sometimes specified. Specifications often state that if concreting operations are suspended for more than 30 minutes, special procedures will be required when the pour is resumed; these include placement of dowels and roughing the exposed surface of the prior pour. This specification is often overly restrictive. A preferable specification would be that an interrupted pour will not require any special procedures as long as the in-place concrete is still plastic. Similarly, many specifications exclude the use of water-reducing admixtures or retarders. There is seldom reason for this restriction. For hot weather concreting, contractors should be responsible for providing mix designs in accordance with ACI 318 and concrete in accordance with ASTM C94, without further, specific restrictions. Why not permit the contractor to use a mix design which includes retarders to permit him to extend the time during which the in-place concrete remains plastic and the batch remains workable? Some specifications concerning cold weather concreting also seem overly restrictive. Temperatures in the ground are satisfactory for proper curing, except for at the surface of the concrete. The primary specification for cold weather concreting should, therefore, relate to any need for surface protection when temperatures are below freezing.

PRE-BID CONFERENCES

Even with the efforts of the geotechnical engineer to provide a thorough subsurface exploration and participate in preparation of plans and specifications, it is inevitable that the drilled pier contractor bidding the project will have questions and require clarification of some aspects of the plans and specifications. A pre-bid meeting, attended by the drilled pier contractors, the geotechnical engineer, the A/E, the general contractor and owner representatives, would be advisable. The drilled pier contractors should be given every opportunity to review the subsurface exploration report and to inspect all rock core obtained during the exploration. When unusual or questionable conditions exist, a drilled pier contractor should install one or more test piers at the site, with the costs borne by the owner. The intent would be to establish a correlation between the equipment and tools proposed for the job and the levels to which shafts can be constructed (compared to anticipated bearing levels), and to resolve questions related to groundwater control and belling.

A secondary but very important function of a pre-bid meeting would be to establish a relationship of mutual respect between the geotechnical engineer, the A/E-owner team and the drilled pier contractors. The intent should be that once the drilled pier contractor is selected, he becomes an integral member of the construction team. This relationship will foster an environment in which all parties work together toward completion of the project, satisfying the design intent and with a minimum of delays and extras.

CONSTRUCTION AND ENGINEERING MONITORING

The geotechnical firm that provided predesign services and consultation on development of the plans and specifications should be the firm that will provide construction monitoring services. It is critical that a good relationship be established between the construction team and the geotechnical firm at the outset of the job. A pre-construction meeting should be held to resolve questions regarding the on-site engineer's role, including determination and documentation of pay quantities and tolerances of the constructed piers. Such a meeting is advisable even though a pre-bid meeting was held. Safety requirements should also be clearly defined prior to construction.

Monitoring of drilled pier construction takes considerable judgment and experience. Therefore, an engineer or geologist should typically be selected to perform this task. If the on-site engineer or geologist is relatively inexperienced, a geotechnical engineer with considerable on-the-job experience with drilled piers must spend time at the project during start-up to provide guidance and instruction.

The on-site engineer or geologist should be thoroughly familiar with the subsurface exploration report and the plans and specs. He should also be trained to anticipate the type and condition of the rock required to satisfy the design. The field evaluation should begin with knowledge of the depth of the shaft and consistency of the materials through which it penetrated. The procedure should include careful observation of the rock sides exposed below the bottom of the casing (if casing is not on bottom) and the bottom rock, and probing of the test hole. If a pier has a bottom elevation far above that anticipated, additional test holes may be warranted. With these data, the on-site engineer must make qualitative and quantitative judgments for approval of the bottom or recommendations for additional work. In addition to assigning an end-bearing capacity to the rock, the engineer may judgmentally allow side shear values as well. A shaft capacity can then be calculated, and the portion of load carried in end-bearing correspondingly reduced.

The judgment for approval or recommendations for additional work should be made by the on-site engineer without undue delay, typically in less than 15 minutes. The engineer should have available various charts that simplify calculations so that his evaluations can be made rapidly. The engineer should inform the drilling contractor's superintendent of any additional work required on the pier in question; this should be done in the presence of the "holemen" who will actually do the work. The engineer will then write these instructions in his field book and on the individual pier evaluation form. A copy of this form should either be presented to the general contractor and drilling subcontractor so they will have a record of the status of each of the piers being worked, or left on a clipboard in the contractor's trailer for their convenient reference.

Once the recommended additional work has been completed, the engineer will make a re-evaluation, including a measurement of the new depth and/or base diameter to calculate the volume of rock removed. The hole will be descended and evaluated as previously, to confirm that the recommended work was completed to satisfaction.

The on-site engineer should maintain, on a daily basis, up-to-date records of observations and the approved bearing elevation and bearing pressure for each pier. These data should be systematically plotted on a foundation plan and comparisons made with subsurface (boring) data. This will allow the engineer to develop a "calibration" between anticipated depths and bearing pressures and the actual constructed conditions.

Most problems with drilled piers relate to improper concreting procedures rather than failure of the foundation material. The monitoring of concreting operations should be an integral part of the on-site engineer's responsibilities. Prior to concreting, the engineer should confirm that loose materials have been removed and that no more than 2 inches (5 cm) of water remains on the bottom. Upon approval, the pump should be removed simultaneously with concrete placement. For large diameter holes where groundwater inflow is heavy, several trucks should pour simultaneously. Alternatively, if water removal methods cannot maintain less than 2 inches (5 cm) on the bottom, tremie methods should be required.

Concrete should be directed down the center of the hole to avoid segregation from hitting the reinforcing bars. The temporary casing should not be pulled until a considerable amount of concrete is placed; the concrete should be maintained several feet above the water level outside the casing until concrete is within about 10 feet (3 m) of the top. If blow-ins have occurred and large voids behind the casing are anticipated, a larger head of concrete should be used. When the final casing is pulled, the concrete surface may become contaminated with water or fall-in soils. It is often necessary for a laborer to descend the hole and pump out or remove the contaminated material. Vibration of high slump concrete is generally not necessary except for within the upper several feet.

SUMMARY

Drilled piers, extended down to hard rock, are excellent building foundations for the support of heavy column loads in the Piedmont. The writer has several recommendations which can make drilled pier design in the Piedmont more economical. If implemented, these recommendations will also better ensure that large unexpected costs and construction-related conflicts relative to pier construction do not occur.

1. An experienced geotechnical firm should be selected by the owner based on qualifications and commitment to the project. The scope of work should be negotiated with the owner-A/E team after selection.

2. Borings should be made at typical spacings of 60 to 100 feet (18 to 30 m). At least five borings should typically be cored to at least 10 feet (3 m). Field work should be performed by drillers with experience in coring Piedmont rock. Accurate subsurface data, particularly related to rock coring, should be clearly submitted in a report, along with cross-sections and estimated contours of rock elevations.

3. The geotechnical engineer should work closely with the A/E team in developing appropriate drilled pier designs and associated specifications. This is considered essential.

4. A pre-bid meeting should be held with prospective drilled pier contractors. The selected contractor should submit details of his proposed equipment, tools and personnel for executing the job; this proposal should be evaluated by the A/E and general contractor. A pre-construction meeting should be held to further resolve any questions.

5. Full-time monitoring of pier construction by the geotechnical engineer is required; close monitoring of concreting operations by experienced personnel is essential. Excess water in the bottom or poor concreting practices can result in serious problems and must be avoided.

CLOSURE

Most of this paper covers hard-rock end-bearing piers. It is recognized that in some areas of the Piedmont, considerable side shear is utilized in design and pier drilling is terminated above rock auger refusal. Associated with this design are construction and contractor conflicts not specifically addressed by this paper. These conflicts can be mitigated by open communications between the geotechnical engineer, AE designers and contractor, with a resultant specification fair to all parties.

Many of the recommendations presented in this paper urge greater involvement by the geotechnical engineer in the design-construction process. The primary beneficiaries of these greater involvements will be the owner, general contractor and drilled pier subcontractor. Because of liability concerns, the geotechnical engineer has resisted, and will continue to resist, involvements which could place greater liability risk on him without a comparable amount of potential reward. For the geotechnical engineer to participate in the ways recommended in this paper, he must be sheltered (held harmless) from added liability. The geotechnical engineer must be under direct contract with the owner and be held harmless by him unless he is negligent in the services provided.

References

1. ADSC: An International Association of Foundation Drilling Contractors, Industry Guidelines for Specification and Inspection of Drilled Shafts, November, 1983, 8 p.

2. D'Appolonia, E. D., J. D'Appolonia, and R. D. Ellison, "Drilled Piers", Chapter 20 from Winterkorn, H. F. and H. Y. Fang, Foundation Engineering Handbook, Van- Nostrand Reinhold Company, New York, 1975, pp. 601-615.

3. Reese, L. C. and Wright, Stephen J., Drilled Shaft Manual, Vol. 1 - Construction Procedures and Design for Axial Loading, U. S. Department of Transportation, Offices of Research and Development, Implementation Division HDV-22, Washington, D.C., July, 1977, 140 p..

4. Kulhawy, Fred H., and Goodman, Richard E., Design of Foundations on Discontinuous Rock, International Conference on Structural Foundations on Rock, Sydney, May, 1980, 12 p.

5. Litke, Scot, "Mandatory Mediation Arbitration Document Approved", Foundation Drilling, ADSC, September/October 1986, pp. 14-17.

6. Standard Specification for the Construction of End Bearing Drilled Piers (ACI 336.1-79), ACI Committee 336, Clyde N. Baker, Jr., Chairman, revised 1985, 9 p.

7. Standards and Specifications for the Drilled Shaft Industry, Association of Drilled Shaft Contractors, October 22, 1976, Revised May 1, 1980 -Updated August, 1983, 18 p.

8. Suggested Design and Construction Procedures for Pier Foundations, ACI Committee 336, ACI 336.3R-72, reaffirmed 1980, 20 p.

9. Wilson, L. Edward, "Soil Engineering: Another Perspective", Foundation Drilling, ADSC, November, 1985, pp. 8-13.

10. Woodward, R. J., W. S. Gardner, and D. M. Greer, Drilled Pier Foundations, McGraw-Hill Book Company, New York, 1972, 288 p.

11. Reese, L. C., "Design and Construction of Drilled Shafts", The Twelfth Terzaghi Lecture, Journal of the Geotechnical Engineering Division, Vol. 104, No. GT1, January, 1978, pp. 91-116.

SUBJECT INDEX
Page number refers to first page of paper.

Construction costs, 87
Construction methods, 87
Contractors, 87
Cost estimates, 15

Drilled pier foundations, 62
Drilled piers, 87

Excavation, 15

Footings, 37
Foundation design, 62

Injection, 37

Piles, 37
Pressuremeter tests, 1

Residual soils, 1, 37
Rocks, 15, 62

Site investigation, 15
Soil modulus, 1
Soil settlement, 1

Weathering, 15, 62

AUTHOR INDEX

Page number refers to first page of paper.

Gardner, William S., 62

Martin, Ray E., 1

Neely, William J., 37

Richardson, Thomas L., 15

Schnabel, James J., 37
Schwartz, Stanley A., 87

Waitkus, Robert A., 37
White, Robert M., 15

RAYMOND H. FOGLER LIBRARY

DATE DUE

BOOKS ARE SUBJECT TO RECALL AFTER TWO WEEKS